This book is due for return or renewal on or before the latest date below

11560

A Geometric Analysis of the Platonic Solids and Other Semi-Regular Polyhedra

With an Introduction to the Phi Ratio

For Teachers, Researchers and the Generally Curious

By Kenneth J.M. MacLean

Book #1 in the Geometric Explorations Series

A Geometric Analysis of the Platonic Solids and Other Semi-Regular Polyhedra: With an Introduction to the Phi Ratio

Another book from **The Big Picture**

Library of Congress Cataloging-in-Publication Data

MacLean, Kenneth James Michael, 1951-
   A geometric analysis of the platonic solids and other semi-regular polyhedra :
with an introduction to the phi ratio : for teachers, researchers and the gener-
ally curious / by Kenneth James Michael MacLean.
      p. cm. -- (Geometric explorations series ; 1)
   Includes bibliographical references and index.
   ISBN-13: 978-1-932690-99-6 (hardcover : alk. paper)
   ISBN-10: 1-932690-99-9 (hardcover : alk. paper)
  1. Polyhedra--Models. 2. Geometry--Study and teaching. 3. Golden section. I.
Title.
   QA491.M33 2007
   516'.156--dc22   2007000763

The Big Picture is an Imprint of Loving Healing Press

Dedication

This book is dedicated to those who can appreciate the logic of numbers and the beauty of nature, for they are both aspects of the same unifying principle.

Acknowledgements

A special acknowledgement goes to Rick Parris for his excellent free software, WinGeom, with which the drawings in this book were made. You can find Rick's programs at http://math.exeter.edu/rparris/

Cover art was provided by George W. Hart. Visit the *Encyclopedia of Polyhedra* at http://www.georgehart.com/virtual-polyhedra/vp.html

Typesetting for the print edition was completed by Victor R. Volkman (Loving Healing Press).

Table of Contents

Introduction

In this book we present a detailed analysis of the five Regular Solids, along with five other important polyhedra:

- The Cube Octahedron
- The Rhombic Dodecahedron
- The Rhombic Triacontahedron
- The Icosa Dodecahedron
- The Star Tetrahedron

We also thoroughly investigate the pentagon, which is nature's key to unlocking the secrets of the Platonic Solids and other important polyhedra. The pentagon is composed entirely of Phi relationships. It turns out that knowledge of Phi is indispensable in the understanding of these polyhedra. This book explains the geometric basis of the Phi ratio and the Fibonacci series, and how they are derived mathematically. For those readers who are serious about sacred geometry, these concepts are essential.

For each polyhedron we calculate:

- Volume
- Surface area
- Central angles to all vertices
- Surface angles
- Dihedral angle
- Centroid to vertex
- Centroid to mid–edge
- Centroid to mid–face
- Side to radius
- Internal planes and their relationships

Volume and surface area is calculated in terms of the edge length of each polyhedron, and also, for comparison purposes, within the unit sphere of radius 1. This sphere is the sphere that contains the outermost vertices of the polyhedron. For the 5 regular solids, all vertices lie on the unit sphere.

These two calculations help us to visualize the polyhedra in relation to each other.

Each polyhedron has an accompanying reference chart.

Appendix A combines the data for all of the polyhedra into one large chart.

A note on dihedral angles:

The dihedral angle is the intersection of two planes, like so:

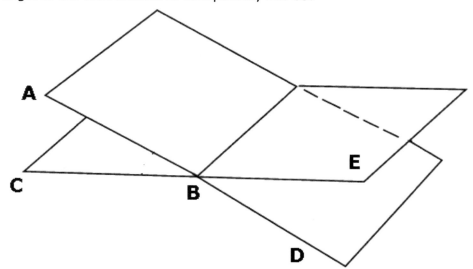

The dihedral angle is usually considered to be the angle ABC. In this book, however, we will use the angle ABE (or CBD), because it is easier to see when building 3 dimensional models. I encourage all readers to build these polyhedra with the Zometool, for it will lead to much greater comprehension of the material.

In this book we avoid trigonometry unless calculating central and dihedral angles. Geometry comes from the words "geo," or earth, and "metrien," to measure. Geometry is about the real and the tangible. It is tempting, when one knows an angle and a side length, to simply use trig to get the other side. One may use the Law of Cosines and other tricks to quickly get distances. Such reasoning does not dig deeply enough, however, and often obscures the geometric detail. In my mind, geometry is all about discovering relationships. In this book I have built models of every solid I have analyzed, allowing me to delve very deeply into the ratio and proportion of these polyhedra. I have found that if one studies long enough, one can always come up with quantities using only right triangles and the Pythagorean Theorem. Such an approach may seem long–winded and unnecessary to some, but it generates a tremendous amount of useful data, and a ton of learning. As researchers we can appreciate the data and as teachers we are always looking for ways to keep mathematics interesting and, dare I say, even exciting for our students.

Make no mistake, there is a lot of exciting stuff in this book for those who can appreciate the beauty of numbers. These polyhedra are merely reflections of Nature herself, and a study of them provides insight into the way the world is structured. Nature is not only beautiful, but highly intelligent. As we explore the polyhedra in this book, this will become apparent over and over again. So crack open the book and get started on a stimulating adventure into the world of geometry!

Part I – Simple Platonic Solids

We begin in Part One with the Tetrahedron, the Octahedron, and the Cube. These polyhedra are all based on $\sqrt{2}$ and $\sqrt{3}$, unlike their much more complex brethren, the Icosahedron and the Dodecahedron, which are based on $\sqrt{5}$ and the Phi ratio.

Chapter 1 – The Tetrahedron

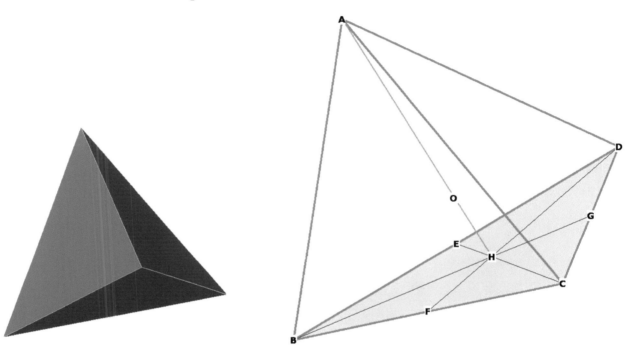

Fig. 1-1: The Tetrahedron

Each side of the tetrahedron is in green. We will refer to the side or edge of the tetrahedron as 'ts.' The tetrahedron has 6 sides, 4 faces and 4 vertices. In Fig. 1-1, the base is marked out in gray: the triangle BCD. Each of the faces is an equilateral triangle.

From The *Equilateral Triangle* (Appendix B) we know that: Area BCD = $\dfrac{\sqrt{3}}{4}\,ts^2$.

Now we need to get the height of the tetrahedron, AH.

From The Equilateral Triangle (Appendix B) we know that: BH = $\dfrac{\frac{1}{\sqrt{3}}\,ts}{}$.
Now that we have BH, we can find AH, the height of the tetrahedron. We will call that h.

$$h^2 = \overrightarrow{AB}^2 - \overrightarrow{HB}^2 = 1ts - \left(\frac{1}{\sqrt{3}}\,ts\right)^2 = \frac{2}{3}\,ts^2.$$

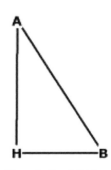

Fig. 1-2: ΔBH

$$h = \frac{\sqrt{2}}{\sqrt{3}} \, ts = 0.816496581 \; ts$$

$$V_{tetrahedron} = 1/3 * \text{area } BCD * h = \frac{1}{3} * \frac{\sqrt{3}}{4} \, ts^2 * \frac{\sqrt{2}}{\sqrt{3}} \, ts = \frac{\sqrt{2}}{12} \, ts^3 = \frac{1}{6\sqrt{2}} \, ts^3 = 0.11785113 \; ts^3$$

What is the surface area of the tetrahedron? It is just the sum of the areas of its 4 faces.

We know from above that the area of a face is $\frac{\sqrt{3}}{4} \, ts^2$.

The total surface area of the tetrahedron $= 4 * \frac{\sqrt{3}}{4} \, ts^2 = \sqrt{3} \; ts^2$.

We have found the volume of the tetrahedron in relation to its side.

Since all 4 vertices of the tetrahedron will fit inside a sphere, what is the relationship of the side of the tetrahedron to the radius of the enclosing sphere?

It's easier to see the radius of the enclosing sphere if we place the tetrahedron inside a cube:

The 4 vertices of the tetrahedron are H,F,C,A.

The 4 faces of the tetrahedron in this picture are: CFH, CFA, HFA, and at the back, HCA.

(The vertex D is a part of the cube, not the tetrahedron).

The vertices of the cube all touch the surface of the sphere. The diameter of the sphere is the diagonal of the cube FD.

In this analysis and the ones following, we will assume that all of our polyhedra are enclosed within a sphere of radius = 1. That way we will be able to accurately assess the relationship between all of the 5 regular solids. The radius of this sphere is OF = OD = 1. How does the side of the tetrahedron relate to the radius of the sphere?

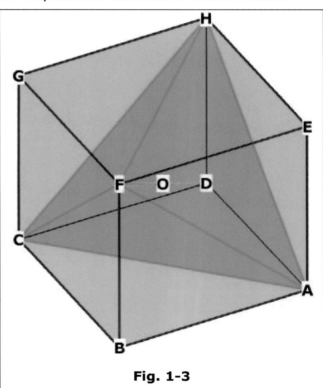

Fig. 1-3

The side of the tetrahedron is the diagonal of the cube face, as can be seen in Figs. 1-3, 1-4.

FC is the side of the tetrahedron, DC is the side of the cube, FD is the diagonal of the cube and the diameter of the sphere enclosing the cube and the tetrahedron.

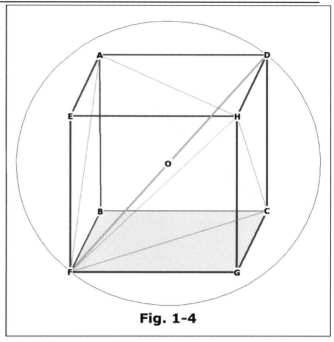

Fig. 1-4

By the Pythagorean Theorem, $\overrightarrow{FD^2} = \overrightarrow{FC^2} + \overrightarrow{DC^2}$.

$FC = \sqrt{2} * FG$, $DC = 1$.

$\overrightarrow{FD^2} = 2 + 1 = 3,$

$FD = \sqrt{3}$ s, where s = side of cube, and $OF = \dfrac{\sqrt{3}}{2}$ s

So FG(side of cube), FC(side of tetrahedron), and FD(diameter of enclosing sphere) have a 1, $\sqrt{2}$, $\sqrt{3}$ relationship.

What is the relationship between the side of the tetrahedron and the radius of the enclosing sphere? We want to know the relationship between OF and FC.

$\dfrac{FC}{FD} = \dfrac{\sqrt{2}}{\sqrt{3}}$. FD = 2 * OF, so $\dfrac{FC}{OF} = \dfrac{2\sqrt{2}}{\sqrt{3}}$. Therefore, $\dfrac{ts}{r} = \dfrac{2\sqrt{2}}{\sqrt{3}} = 1.632993162$.

So we write $ts = \dfrac{2\sqrt{2}}{\sqrt{3}}$ r, $r = \dfrac{\sqrt{3}}{2\sqrt{2}}$ ts .

Where is the center of mass of the tetrahedron? We have marked it as O in the preceding figures. The centroid is just the distance O to any of the tetrahedron vertices, or the radius of the enclosing sphere. The center of mass of the cube is also the center of mass of the tetrahedron. As is evident from Figure 3 and Figure 7, every vertex of the tetrahedron is a vertex of the cube.

We already have OF, or r, which is $\dfrac{\sqrt{3}}{2\sqrt{2}}$ ts . If we compare this distance to the height of the tetrahe-

dron, we get $\dfrac{r}{h} = \dfrac{\left(\dfrac{\sqrt{3}}{2\sqrt{2}} ts\right)}{\left(\dfrac{\sqrt{2}}{\sqrt{3}} ts\right)} = \dfrac{\sqrt{3}}{2\sqrt{2}} \times \dfrac{\sqrt{3}}{\sqrt{2}} = \dfrac{3}{4} = 0.75$

Therefore AO = 3 x OH.

Compare this to the equilateral triangle, where the relationship is 2 to 1.

What is the central angle of the tetrahedron?

The central angle is the angle from one vertex, through the centroid O, to another vertex. This can best be seen from Figure 1-3.

We calculate everything in terms of the side of the tetrahedron.

OF = OH = radius of enclosing sphere.

FH = ts.

$$IF = \frac{1}{2}FH = \frac{1}{2}ts$$

$$\sin\angle IOF = \frac{IF}{OF} = \frac{1}{2} \times \frac{2\sqrt{2}}{\sqrt{3}} = \sin\left(\frac{\sqrt{2}}{\sqrt{3}}\right)$$

$$\angle IOF = \arcsin\left(\frac{\sqrt{2}}{\sqrt{3}}\right) = 54.73561032°$$

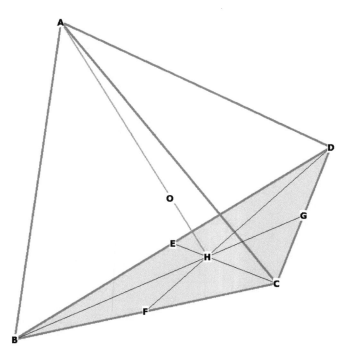

Fig. 1-1: repeated.

The central angle FOH = 2 x ∠IOF.

Central angle = 109.47122064°

What is the surface angle of the tetrahedron? The surface angle is that angle made by the angles within the faces. Since each face of the tetrahedron is an equilateral triangle, the surface angle = 60°

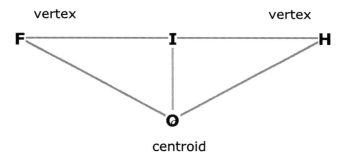

Fig. 1-5: Central Angle FOH

(From Fig. 1-7)

What is the dihedral angle of the tetrahedron?

The dihedral angle is the angle formed by the intersection of 2 planes:

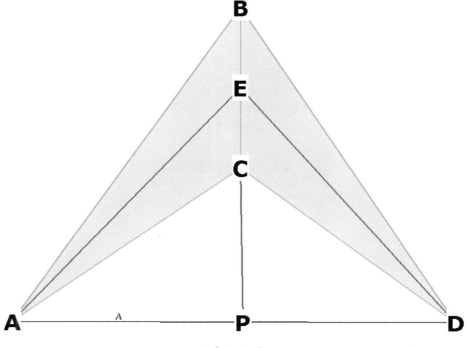

Fig. 1-6

BAC and BDC are 2 intersecting faces of the tetrahedron. E is at the midpoint of BC.

AE and DE are lines that go through the middle of each face and hit E.

The dihedral angle is $\angle$ AED.

The triangle EPD is right. $\angle$ PED = 1/2 $\angle$ AED by construction.

AD is the side of the tetrahedron s, so PD = ½(ts)

ED, the height of the tetrahedron face = $\dfrac{\sqrt{3}}{2}$ ts .

$$\sin(\angle \text{PED}) = \text{PD/ED} = \dfrac{\dfrac{1}{2}}{\dfrac{\sqrt{3}}{2}} = \dfrac{1}{2} \times \dfrac{2}{\sqrt{3}} = \dfrac{1}{\sqrt{3}} .$$

$\angle$ PED = arcsin ($\dfrac{1}{\sqrt{3}}$) = 35.26438968°

$\angle$ AED = 2 * $\angle$ PED, Dihedral angle = 70.52877936°

Now we will calculate the distances between the centroid and

1) the middle of one of the faces (OJ)
2) the midpoint of one of the sides (OI)
3) any vertex (OH)

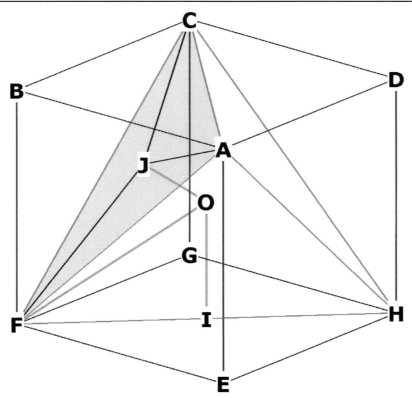

Fig. 1-7: Centroid Distances

OF = distance from centroid to a vertex, in this case, F,

OI = distance from centroid to middle of a side, in this case, the side FH

OJ = distance from centroid to middle of a face, in this case, face ACF,

The 4 faces are CFH, CFA, HFA and HCA as before. The middle of face CFA is J. The midpoint of the side of the tetra FH is at I. We have already figured out OF = OC = OA = OH, the distance from the centroid to any vertex. This is, as you recall, $\dfrac{\sqrt{3}}{2\sqrt{2}}$ ts .

Let's get OI first.

$$\overline{OI^2} = \overline{OF^2} - \overline{IF^2} = \left(\frac{\sqrt{3}}{2\sqrt{2}}\,ts\right)^2 - \left(\frac{1}{2}\,ts\right)^2 = \frac{3}{8}\,ts - \frac{1}{4}\,ts = \frac{1}{8}\,ts$$

$$OI = \frac{1}{2\sqrt{2}}\,ts.$$

The triangle FJO in Fig. 1-7 is right because ∡ OJF is right[1]. This will enable us to get the distance OJ.

_____

[1] Note to teachers: if you build an octahedron and a rhombic dodecahedron off the octahedron faces (use the Zometool), and then attach a tetrahedron off one of the octahedron faces, the centroid of that tetrahedron will be the vertex of the rhombic dodecahedron. In fact, the distance from the centroid of the tetrahedron to each of its vertices is the edge length of the rhombic dodecahedron! Nature is not only beautiful, but also logical and consistent.

From Equilateral Triangle we know that $FJ = \dfrac{1}{\sqrt{3}} \, ts$.

$$\overrightarrow{OJ}^2 = \overrightarrow{OF}^2 - \overrightarrow{FJ}^2 = \left(\frac{\sqrt{3}}{2\sqrt{2}} \, ts\right)^2 - \left(\frac{1}{\sqrt{3}} \, ts\right)^2 = \frac{3}{8} \, ts^2 - \frac{1}{3} \, ts^2 = \frac{1}{24} \, ts^2 \, ,$$

$$OJ = \frac{1}{2\sqrt{6}} \, ts \, .$$

To get a good idea of relative distances,

re-write OF as $\dfrac{3}{2\sqrt{6}} \, ts$ and OI as $\dfrac{\sqrt{3}}{2\sqrt{6}} \, ts$.

Then the relationship between OJ, OI and OF is 1, $\sqrt{3}$, 3.

Tetrahedron Reference Tables

Volume in terms of s	Volume in Unit Sphere	Surface Area in terms of s	Surface Area in Unit Sphere
0.11785113 s³	0.513200238 r³	1.732050808 s²	4.618802155 r²

Central Angle:	Dihedral Angle:	Surface Angle:
109.47122064°	70.52877936°	60°

Centroid To: Vertex	Centroid To: Mid−Edge	Centroid To: Mid−Face
1.0 r	0.577350269 r	0.33333333 r
0.612372436 s	0.353553391 s	0.204124145 s

Side / radius
1.632993162

Chapter 2 – The Octahedron

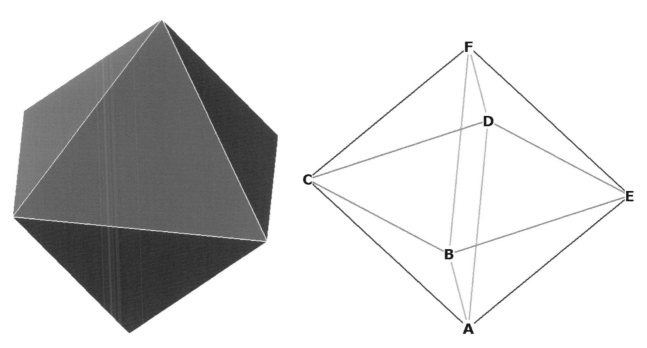

Fig. 2-1: showing the 8 faces and 3 squares of the octahedron

The Octahedron has 12 sides, 8 faces and 6 vertices. Count them!

Each of the octahedron's 8 faces is an equilateral triangle, just like the tetrahedron, but the tetrahedron only has 4 faces.

Notice how the octahedron can be considered to be formed from 3 orthogonal squares:

The square BCDE, the square ABFD, and the square ACFE, all 3 of which are planes and all 3 of which are perpendicular to each other.

The octahedron's face angles are all equal because they are all 60°, however, the internal angles of the squares are all 90°. I guess it depends on which way you want to look at it! That's spatial geometry: the position of the observer relative to an object can yield quite different perspectives.

Let's calculate the volume of the octahedron. All of the octahedron's vertices touch upon the surface of the same sphere that encloses the tetrahedron. The centroid is at O. We will place a midpoint G in the middle of the face CDF. We also place a midpoint H on CD, one of the sides of the octahedron. The side or edge of the octahedron will be hereinafter referred to as 'os.'

What is the volume of an octahedron?

The octahedron consists of 2 pyramids, face-bonded. One pyramid is at A-BCDE, the other at F-BCDE.

The base of each pyramid is the square BCDE.

The area of the base is then just os x os = os².

The height of the pyramids are h = OF = OA.

But OF = OA = OC, since the octahedron is composed of 3 squares. FC is just a side of the octahedron.

Therefore we can write:

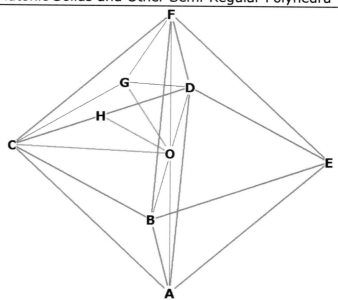

Fig. 2-2

$$OF^2 + OC^2 = FC^2, \quad OF^2 + OF^2 = FC^2, \quad 2OF^2 = os^2.$$

$$OF = h = \frac{1}{\sqrt{2}} \, os$$

So $V_{1 \, pyramid}$ = 1/3 × area of base × height = 1/3 $(os^2)\dfrac{1}{\sqrt{2}} os$ = $\dfrac{1}{3\sqrt{2}} os^3$.

$$V_{octahedron} = 2 \times V_{1 \, pyramid} = \frac{2}{3\sqrt{2}} os^3,$$

$$V_{octahedron} = \frac{\sqrt{2}}{3} os^3 = 0.471404521 \; os^3$$

What is the surface area of the octahedron? It is just the sum of the area of the faces. Since each face is an equilateral triangle, we know from The Equilateral Triangle that the area for one face is $\dfrac{\sqrt{3}}{4} os^2$.

$$\text{Surface area (octa)} = 8 * \frac{\sqrt{3}}{4} os^2 = 2\sqrt{3} \; os^2.$$

What is the relationship between the radius of the enclosing sphere and the side of the octahedron?

All 6 vertices of the octahedron touch the surface of the sphere. Therefore the diameter of the sphere is just FA = CE = DB.

The radius is 1/2 that, or h, which we found above to be $\dfrac{1}{\sqrt{2}} os$.

So $r = \dfrac{1}{\sqrt{2}} os$. os = $\sqrt{2}r$.

The centroid of the octahedron is at O.

What is the central angle of the octahedron? 90°.

Since the octahedron is made of 3 orthogonal squares, the angle through O from any 2 adjacent vertices must be 90°.   To see this, look at ∡FOC.

What are the surface angles of the octahedron? 60°, because each of the faces is an equilateral triangle.

What is the dihedral angle of the octahedron?

The dihedral angle is the angle formed by the intersection of 2 planes:

AXF is the dihedral angle because it is the intersection of the planes ABC and FBC.

ACB and FBC are faces of the octahedron. AX and XD are lines from the face vertex through the mid-face and bisecting an edge of the octahedron, BC.

AF is the diameter of the enclosing sphere, and AO and OF = r = h.

OX is just 1/2 of a side of the octahedron BC. You can see this marked in Fig. 2-2- as the line

OH. So OX = $\frac{1}{2}$ os .

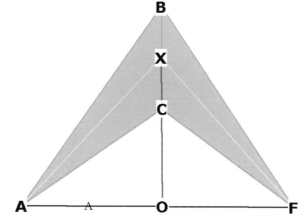

Fig. 2-3: Dihedral Angle as an Intersection of 2 Planes

Triangle XOF is right.

$$\tan(\angle OXF) = \frac{OF}{OX} = \frac{\frac{1}{\sqrt{2}}}{\frac{1}{2}} = \frac{2}{\sqrt{2}} = \sqrt{2}.$$

$\angle OXF = \arctan(\sqrt{2}) = 54.73561032° = \frac{1}{2}\angle AXF$

Dihedral angle = 109.4712206°

What is the distance from the centroid to the midpoint of one of the sides?

The distance from the centroid to the midpoint of a side can be seen in Fig. 2-2 as OH.

We already know this to be $\left(\frac{1}{2}\right)$os .

What is the distance from the centroid to the middle of one of the faces?

The distance from the centroid to the middle of a face can be seen in Fig. 2-2 as OG.

Triangle FGO (see Fig. 1-2) is right, because OG is perpendicular to the face FBC by construction

From The Equilateral Triangle we know that $FG = \dfrac{1}{\sqrt{3}}os$.

$OF = h = \dfrac{1}{\sqrt{2}}os$

So $\overline{OG}^2 = \overline{OF}^2 - \overline{FG}^2 = \dfrac{1}{2}os^2 - \dfrac{1}{3}os^2 = \dfrac{1}{6}os^2$

$OG = \dfrac{1}{\sqrt{6}}os$

The distance from the centroid O to the middle of one of the faces $= \dfrac{1}{\sqrt{6}}os$.

Fig. 2-4: Right Triangle FGO

What is the distance from the centroid to any vertex?

It is just the radius of the enclosing sphere, or $\dfrac{1}{\sqrt{2}}os$.

Octahedron Reference Tables

Volume in terms of s	Volume in Unit Sphere	Surface Area in terms of s	Surface Area in Unit Sphere
0.471404521 s³	1.333333 r³	3.464101615 s²	6.92820323 r²

Central Angle:	Dihedral Angle:	Surface Angle:
90°	109.4712206°	60°

Centroid To: Vertex	Centroid To: Mid−Edge	Centroid To: Mid−Face
1.0 r	0.707106781 r	0.577350269 r
0.707106781 s	0.5 s	0.408248291 s

Side / radius
1.414213562

Chapter 3 – The Cube

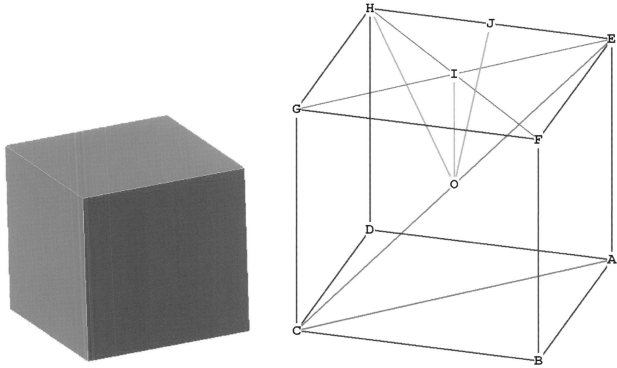

Fig. 3-1: The Cube

The cube has 12 sides, 8 vertices and 6 faces. All of the vertices of the cube lie on the surface of the sphere. So the diameter of the enclosing sphere is CE, or any diagonal of the cube that goes through the centroid O.

The radius of the sphere is 1/2 that; OE, for example.

In other words, the radius is any line from O to any vertex.

The side of the cube will be referred to as 's.'

The volume of the cube is just s * s * s = s³

The surface area is just 6 faces * area of each face.

The area of each face is s*s = s², so the surface area = 6s².

What is the relationship between the radius of the enclosing sphere and the side of the cube?

Look at the right triangle OIE in Fig.3-1. To get the radius OE in terms of the side of the cube, notice that OI is just 1/2 BF, the side of the cube.   So OI $=\dfrac{1}{2}$s .

IE is 1/2 the diagonal of the face EFGH. By the theorem of Pythagoras, GE is s$\sqrt{2}$, because GH and HE are both s.

Therefore IE $=\dfrac{1}{\sqrt{2}}$s .

$$\overline{OE}^2 = r^2 = \overline{OI}^2 + \overline{IE}^2 = \frac{1}{4}s^2 + \frac{1}{2}s^2 = \frac{3}{4}s^2$$

$$OE = \frac{\sqrt{3}}{2}s.$$

$$r = \frac{\sqrt{3}}{2}s, \quad s = \frac{2}{\sqrt{3}}r$$

The volume of the cube in the unit sphere of radius = 1 is therefore, $s \times s \times s = \left(\frac{2}{\sqrt{3}}r\right)^3 = \frac{8}{3\sqrt{3}}r^3.$

What is the central angle of the cube?

The central angle is the angle through O from any two adjacent vertices. So in Fig. 3-1, that would be $\angle HOE$.

We bisect HE at J to get the triangle OJE, which is right by construction.

$\angle JOE$ will be one-half the central angle HOE.

$$\sin(\angle JOE) = JE \,/\, OE = \frac{\frac{1}{2}}{\frac{\sqrt{3}}{2}} = \frac{1}{\sqrt{3}}.$$

$$\angle JOE = \arcsin\left(\frac{1}{\sqrt{3}}\right) = 35.26438968°$$

$\angle HOE = 2 \times \angle JOE = 70.52877936°$

Central angle = 70.52877936°.

Surface angle = 90°.

Dihedral angle = 90°.

You can see this by observing the intersection of planes BCGF and DCGH.

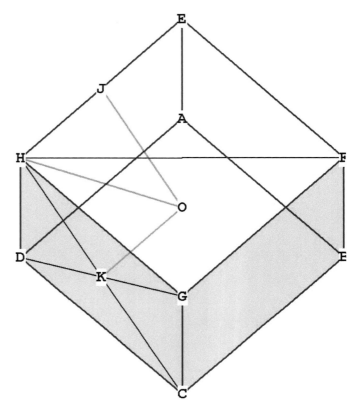

Fig.3-2

What is the distance from the centroid to mid-face? This is $OI = \frac{1}{2}s$

What is the distance from the centroid to mid-edge? This is OJ.

By Pythagorean Theorem we may write

$$\overline{OJ}^2 = OE^2 - \overline{JE}^2 = \left(\frac{\sqrt{3}}{2}s\right)^2 - \frac{1}{4}s^2 = \frac{3}{4}s^2 - \frac{1}{4}s^2 = \frac{1}{2}s^2.$$

$$OJ = \frac{1}{\sqrt{2}}s.$$

What is the distance from the centroid to any vertex? This is $OE = \frac{\sqrt{3}}{2}s.$

Looking at comparative distances, we have .5, .707106781, .866015404.

Cube Reference Tables

Volume in terms of s	Volume in Unit Sphere	Surface Area in terms of s	Surface Area in Unit Sphere
1.0 s³	1.539600718 r³	6.0 s²	8.0 r²

Central Angle:	Dihedral Angle:	Surface Angle:
70.52877936°	90°	90°

Centroid To: Vertex	Centroid To: Mid–Edge	Centroid To: Mid–Face
1.0 r	0.816496581 r	0.577350269 r
0.866025404 s	0.707106781 s	0.5 s

Side / radius
1.154700538

Part II – Phi and the Pentagon
Chapter 4 – The Phi Ratio

In order to properly study the remaining Platonic Solids, it is necessary to learn about the Phi ratio.

It turns out that the dodecahedron and the icosahedron are both based on $\sqrt{5}$ and Phi.

Introduction to the Phi ratio

The Phi Ratio is a proportion. A proportion is a relationship between one quantity and another quantity. It is often very helpful to have something to compare a thing to, in order to establish familiarity with it. For instance, to say "that building is 100" really doesn't help us at all. But if we say "the building is 100 stories" we can get an idea in our minds how large it is. By associating a metric, or measurement, to a number we give it meaning and make it real.

In geometry, comparing 2 quantities to each other helps us establish a relationship between the two quantities. A pure number has no reality, because there is nothing to compare that number to. In geometry, if we can start with a known quantity or length, and then compare every other element in the drawing to that known quantity, we can stay clear on how everything in the drawing relates to everything else.

The Phi Ratio has been known for millennia. It has also been referred to as the Golden Ratio. Euclid called it "Division in Mean and Extreme Ratio." I sometimes shorten this to EMR.

The division into mean and extreme ratio is extremely important because in such a division, there is perfection. The division of any line into EMR, for example, can be continued infinitely small or infinitely large without the slightest error, or "round off" of any kind. So division in EMR or Phi Ratio leads to perfect harmonious growth.

The Phi ratio comes from the division of a line segment such that "The smaller is to the larger as the larger is to the whole."

Consider the line segment GX′, divided at O into line segments GO with smaller length a, and OX′, with larger length b.

$$\text{G} \underline{\hspace{3cm}} \text{O} \underline{\hspace{5cm}} \text{X'}$$
$$\quad\quad\quad a \quad\quad\quad\quad\quad\quad\quad b$$

Mathematically stated, the above statement becomes $\dfrac{a}{b} = \dfrac{b}{a+b}$.

Let b = OX′ = 1.

So $\dfrac{a}{1} = \dfrac{1}{a+1}$. Then $a^2 + a = 1$ and $a^2 + a - 1 = 0$.

$a^2 + a = 1$ is a second degree polynomial, and you can solve it easily on your calculator. Just press the "poly" key and enter 1, 1,-1.

The calculator will show 2 solutions, the first is 0.618033989 $= \dfrac{1}{\Phi}$.

The second is -1.618033989 $= -\Phi$.

Since a is shorter than b, and a is a positive length, a must be $\dfrac{1}{\Phi}$. Then a + b = 1 + $\dfrac{1}{\Phi}$ = Φ.

"The smaller is to the larger:" The ratio b/a = $\dfrac{1}{\left(\dfrac{1}{\Phi}\right)}$ = Φ "As the larger is to the whole:"

$$\dfrac{b + a}{b} = \dfrac{\Phi}{1} = \Phi$$

Although Φ is a number it is more properly stated as a geometric *ratio* of two numbers. A ratio is a relationship between two things. When we investigate the pentagon, we'll see that Phi is stated mathematically as $\left(\dfrac{\sqrt{5}+1}{2}\right)$, and that $\dfrac{1}{\Phi}$ is stated mathematically as $\left(\dfrac{\sqrt{5}-1}{2}\right)$. This might seem uninformative at first, but there is a very understandable and intuitively sensible geometric interpretation of these quantities that you'll be able to easily grasp. Now let's look at the Phi Ratio in action.

The Golden Triangle

The golden triangle is a phi ratio triangle: Triangle ABC is a golden triangle because when either of the long sides (AB or AC) is divided by the short side (BC), the result is Phi (Φ).

You can see how the triangle divides into itself, always making a triangle with sides and angles in the same proportion.

For example, the triangle BCD is also a golden triangle. Triangles CDE, DEF, and EFG are as well, all having the property that the long side of each triangle divided by the short side = Φ.

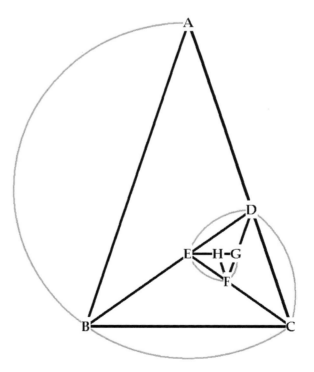

Fig. 4-1: The Golden Triangle

In the large triangle ABC, for example, if the line segments AB and BC were combined to form 1 line, that line would be divided in Phi ratio at B:]

A_____B_____C

Fig. 4-2 The line ABC divided in Phi Ratio at B.

Exercise: Requires paper, compass and straight-edge.

Take the smaller segment of triangle ABC, BC, and transfer that length to the line AC. Do this by pinning the compass at B with length BC to get the distance, then swinging the marker end until it hits AC at D. Now AC has been divided in Phi ratio at the point D! Connect BD with the straight-edge. A second triangle has been now formed, BCD, which is itself a golden mean triangle. The process is absurdly simple, for once a Golden Triangle is constructed, it can be subdivided or grown with hardly any thought whatsoever!

If we let AB = AC = 1, then

$$BC = BD = AD = \frac{1}{\Phi}.$$

$$DC = \frac{1}{\Phi^2}, \text{ because } \frac{1}{\Phi} + \frac{1}{\Phi^2} = 1.$$

Now do the same thing with triangle BCD. Take the short side CD, and transfer that distance onto BD to find the point E. Then connect CE with the straight-edge. Another golden mean rectangle is formed, CDE. Continue in this fashion to subdivide the larger segment of each triangle by the smaller.

You can see that this process can occur indefinitely. The process can be reversed to make the object larger, on to infinity. The only reason this works is that the division into Phi ratio occurs absolutely perfectly, with not even the tiniest error. (Of course there is always construction error, but mathematically the Phi ratio is perfect division). Data:

$$BC = BD = AD = \frac{1}{\Phi}.$$

$$CD = CE = BE = \frac{1}{\Phi^2}.$$

$$DE = DF = CF = \frac{1}{\Phi^3}.$$

$$EF = EG = DG = \frac{1}{\Phi^4}, \text{ etc.}$$

A little trigonometry reveals that these golden triangles have angles 36-72-72. (Do this as an exercise).

Notice the spiral that forms around the triangles. This is called a logarithmic spiral. If you observe nature, you will find that she does not use straight lines, she uses curves.

The Golden Rectangle

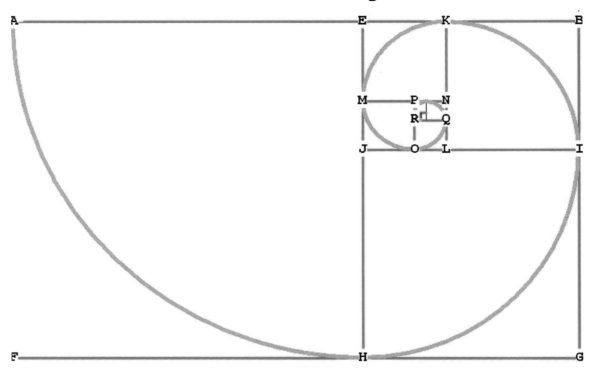

The long sides of the rectangle ABGF (AB and FG) have first been divided into Phi ratio at E and H; meaning that, for example, the ratio FG / FH = Φ, and the ratio FH / GH = Φ. Remember that Phi is a ratio describing the relationship between two quantities.

Exercise:

To make the smaller golden rectangle EBGH inside ABGF, simply pin your compass at A and mark out the distance to F. Then pin your compass at F and mark off the distance to H on the line FG. With the straightedge, draw a perpendicular to the point H. Now EBGH is another golden rectangle. Do the same to make golden rectangle EBIJ, and so on. Notice that AEHF is a square, and so is JIGH, and so on. This process can go on downward or upward indefinitely. Again, it is really simple and painless. Once you get the hang of it, perfect growth or division can occur almost automatically. Notice that the spiral hits all of the points where a line has been divided into Phi ratio: H, where FG has been divided into Phi ratio; I, where GB has been divided into Phi ratio, etc.

What have we learned? That the division into Phi ratio leads to absolutely perfect, harmonious growth. Unfortunately, perfection has no beginning or ending. In other words, we could make our rectangles smaller and smaller, and they would begin to converge at a particular point, but never reach it. As you can see from the above diagram, the spiral keeps circling round and round, never quite reaching the center, or growing outward without end. When nature builds something, she needs a starting and an ending point. Fortunately, hidden within the properties of Phi is the answer.

Chapter 5 – The Fibonacci Series

Because division or growth by the Golden Mean has no beginning or ending, it is not a good candidate for constructing forms of any kind. It is necessary to have a definable starting point if you want to build anything.

However, there is an approximation to the Golden Mean that nature uses, called the Fibonacci Sequence. Leonardo Fibonacci was a monk who noticed that branches on trees, leaves on flowers, and seeds in pine cones and sunflower seeds arranged themselves in this sequence.

The Fibonacci Sequence is based on the golden mean ratio, or Phi (Φ). To the left of the equal sign is Φ raised to a power, and to the right of the equal sign is the Fibonacci Sequence:

$$\Phi^1 = 1\Phi + 0$$
$$\Phi^2 = 1\Phi + 1$$
$$\Phi^3 = 2\Phi + 1$$
$$\Phi^4 = 3\Phi + 2$$
$$\Phi^5 = 5\Phi + 3$$
$$\Phi^6 = 8\Phi + 5$$
$$\Phi^7 = 13\Phi + 8$$
$$\Phi^8 = 21\Phi + 13$$
$$\Phi^9 = 34\Phi + 21$$
$$\Phi^{10} = 55\Phi + 34$$

...

Each digit in the second column to the left of the Φ symbol is the sum of the 2 before it in the previous row:

1 + 0 = 1, 1 + 1 = 2, 2+1 = 3, 3 + 2 = 5, 5 + 3 = 8, ... and so on.

Notice that when one divides a digit by the one before it in the sequence, the result rapidly approaches Φ :

1 / 1 = 1.0

2 / 1 = 2.0

3 / 2 = 1.5

5 / 3 = 1.67

8 / 5 = 1.60

13 / 8 = 1.625

21 / 13 = 1.6153846

34 / 21 = 1.6190476

55 / 34 = 1.617647

...

Here is a chart that shows this:

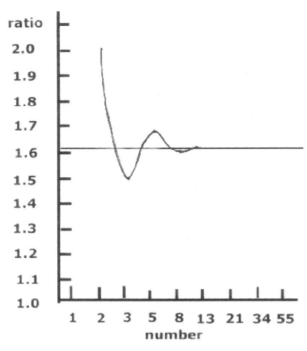

Fig. 5-1: Fibonacci Ratio Converging on Φ

Notice that Φ is approached both from above and below. This is not an asymptotic approach, but one which affords a full view of both sides. Like a signal locking in on the target, the Fibonacci sequence homes in on Φ.

The Φ ratio is attained from the division of integers. Integers, or whole numbers, represent things that can be designed and built in the 'real world.' Even though Φ is unattainable, nature can closely approximate it, as we can see from Figs. 5-2 to 5-4:

Fig. 5-2: A Nautilus shell

Fig. 5-3: Echeveria Agavoides ("Molded Wax Agave")

Fig. 5-4: Helianthus Annus ("Common Sunflower")

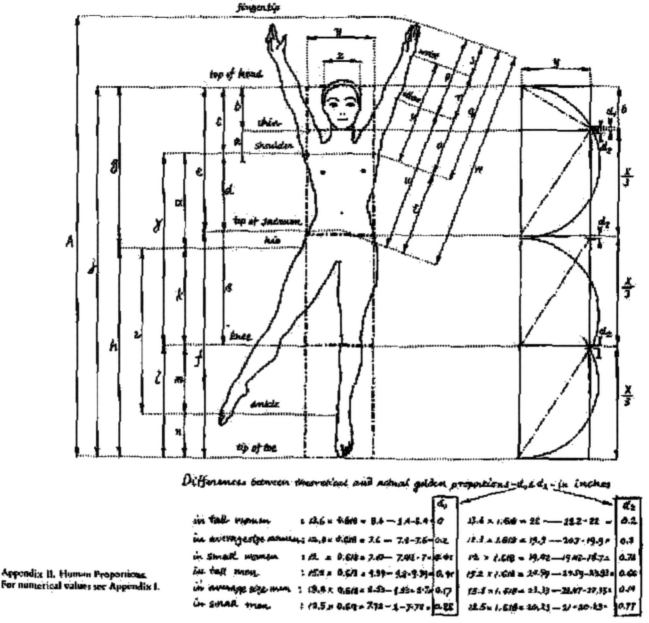

Fig. 5-5: Human Body Φ Relationships[2]

The human body has many Phi/Fibonacci relationships.

- Leg: The distance from the hip to the knee, and from the knee to the ankle.
- Face: The distance from the top of the head to the nose, and from the nose to the chin.
- Arm: The distance between the shoulder joint to the elbow, and from the elbow to the wrist.
- Hand: The distance between the wrist, the knuckles, the first and second joints of the fingers, and the finger tips.

---

[2] From *The Power of Limits* by Gyorgy Doczni, p. 143.

Of course every body is different and these relationships are only approximate. Once you become aware of the Fibonacci sequence / Phi ratio, however, you begin to see it in many life forms.

(As an aside, note that in the Fibonacci sequence, you don't have to start with 1. Begin with any number and proceed with the rule "the current number is the sum of the 2 previous numbers." Do the division of the current number by the prior number in the sequence and Φ will always be the result. For example, begin with 5. Then the sequence goes 5, 5, 10, 15, 25, 40, 65, 105... The ratios between the two numbers will be exactly the same as the sequence progresses.)

Chapter 6 – Properties of the Phi Ratio

Because Φ divides into itself perfectly, it has the following amazing properties:

$$\Phi^{-1} = \frac{1}{\Phi^2} + \frac{1}{\Phi^3} \qquad\qquad \text{OR} \qquad \Phi^{-1} = \Phi^{-2} + \Phi^{-3}$$

$$\Phi^0 = \frac{1}{\Phi} + \frac{1}{\Phi^2} \qquad\qquad \text{OR} \qquad \Phi^0 = \Phi^{-1} + \Phi^{-2} \;,\; \Phi^0 = 1$$

$$\Phi^1 = 1 + \frac{1}{\Phi} \qquad\qquad \text{OR} \qquad \Phi^1 = \Phi^0 + \Phi^{-1}$$

$$\Phi^2 = \Phi + 1 \qquad\qquad \text{OR} \qquad \Phi^2 = \Phi^1 + \Phi^0$$

$$\Phi^3 = \Phi^2 + \Phi \qquad\qquad \text{OR} \qquad \Phi^3 = \Phi^2 + \Phi^1$$

$$\Phi^4 = \Phi^3 + \Phi^2$$

So $\Phi^n = \Phi^{n-1} + \Phi^{n-2}$, n is any integer.

Φ to any power is the sum of the 2 powers before it.

Multiples of Φ can be combined to form any integer.
Here is a brief table:

$$1 = \frac{1}{\Phi} + \frac{1}{\Phi^2} \quad \text{or } \Phi - \frac{1}{\Phi}$$

$$2 = \Phi + \frac{1}{\Phi^2} \quad \text{or } \Phi^2 - \frac{1}{\Phi}$$

$$3 = \Phi^2 + \frac{1}{\Phi^2}$$

$$4 = \Phi^3 - \frac{1}{\Phi^3} \quad \text{or } \Phi^2 + 1 + \frac{1}{\Phi^2}$$

$$5 = \Phi^3 + \frac{1}{\Phi} + \frac{1}{\Phi^4} \quad \text{or } 5\left(\Phi - \frac{1}{\Phi}\right)$$

$$6 = \text{multiple of } 2,3$$

$$7 = \Phi^4 + \frac{1}{\Phi^4}$$

$$8 = \text{multiple of } 2,4$$

$$9 = \text{multiple of } 3$$

$$10 = \text{multiple of } 2 , 5$$

$$11 = \Phi^5 - \frac{1}{\Phi^5} \text{, or } \Phi^4 + \Phi^2 + 1 + \frac{1}{\Phi^2} + \frac{1}{\Phi^4} \quad \text{(notice the symmetry)}$$

$$12 = \text{multiple of } 2, 3, 4$$

$$13 = \Phi^5 + \Phi + \frac{2}{\Phi^4}$$

etc.

Apparently, powers of Phi added or subtracted to their inverses always equals an integer.

Chapter 7 – The Pentagon

In this section we'll learn how to construct the pentagon with compass and straight–edge. In the next section, we'll look at what we've done mathematically. Why are we spending so much time on the pentagon? Because it is fundamental to the construction of both the icosahedron and the do-decahedron. To follow along, get out your compass and straight–edge.

First, begin with a circle. Pick a point anywhere on the circle, say X, and with the straightedge draw a line through O, the center, until it intersects the circle again at X'. The radius of the circle is OX = OX'. Let this distance be 1.

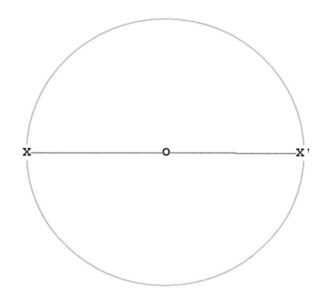

Fig. 7-1: Constructing the Pentagon – Step 1

Next, bisect the line XX' with the compass. Pin compass at X, place the other leg at X' and draw an arc from top (above A) to bottom (below Q).

Pin compass at X', and place the other leg at X. Draw an arc to intersect the arcs you just drew.

I have placed 2 little x's on AQ where both arcs meet. Connect the x's and you have the line AQ in Fig. 7-2 (on following page):

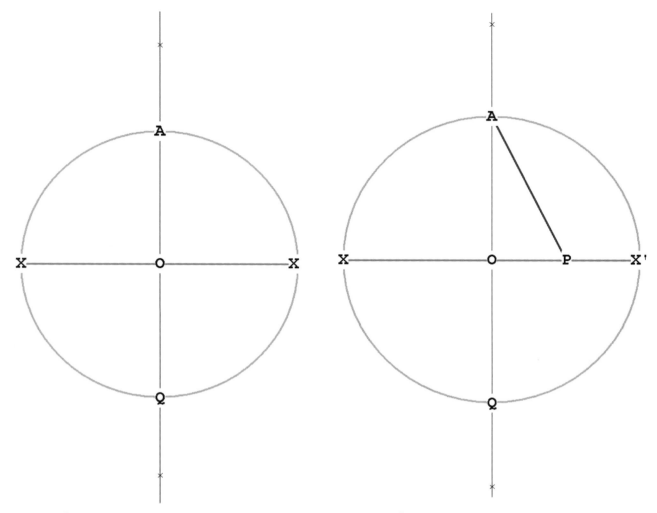

Fig. 7-2: Pentagon – Step 2   **Fig. 7-3: Pentagon – Step 3**

Next, bisect the line OX' at P. We have now divided OX' in half. With the straight-edge connect line AP.

Now take the distance AP and transfer it to XX'. Do this by pinning the compass at P, with length AP, and drawing an arc down to XX'. Mark that point as G. Now AP = PG and the distance AG (marked in green) is the side of the pentagon. (AG is > AP).

Now, pin the compass at A, with length AG

Draw an arc with the compass from AG through to B, and the other way until it hits the side of the circle at C. Now we have three of the 5 points of the pentagon, A, B, and C.

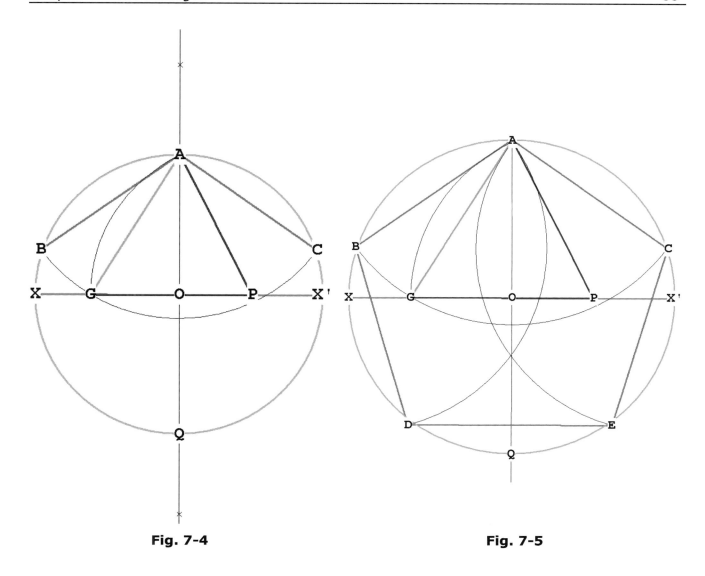

Fig. 7-4   **Fig. 7-5**

Now, walk the compass around the circle by either pinning it at B or C, with distance AB = AC, and finding D and E. Connect up ABDECA and you have the pentagon.

Why is the pentagon so important?

A nucleotide.

Nucleotides are the basic units of DNA. The phosphate group to the left is square shaped and the nitrogenous base is a hexagon. The central structure, the deoxyribose sugar, is pentagonally shaped. Nature uses geometry in the construction of life!

The Composition of the Pentagon

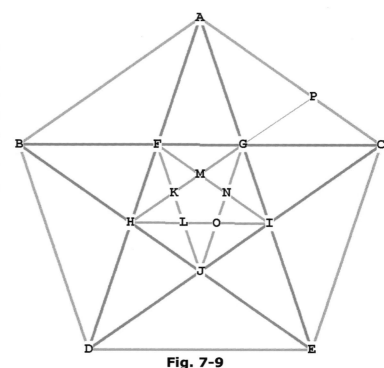

In Fig. 7-9, the diagonals of the pentagon (in green) have been marked. These diagonals form another pentagon, GFHIJ, which is rotated 36 degrees from the original. By connecting the diagonals of GFHIJ, we form another pentagon, KLONM, which is rotated 36 degrees from GFHIJ.

Let the sides of the pentagon be of length 1. Then:

Fig. 7-9

BC = AD = AE = CD = BE = Φ

AB = BC = CD = DE = EA =   1 (the sides of the pentagon)

BF = GC =BH = JE =AF = HD = DJ = IC = AG = IE = $\dfrac{1}{\Phi}$

FG = FH = HJ = JI = GI = $\dfrac{1}{\Phi^2}$

FK = FM = GM = GN = IN = IO = JO = JL = HL = HK = $\dfrac{1}{\Phi^3}$

OL = LK = KM = MN = NO = $\dfrac{1}{\Phi^4}$

etc.

Every one of the diagonals of the pentagon divides the other diagonals in Mean and Extreme (Phi) ratio!

Because the sides and the diagonals of the pentagon are in Phi relationship, every one of the triangles within the pentagon is a Phi ratio triangle. The number of these triangles is too numerous to mention, but they are all either 36–36–108 triangles (like AFB and GIJ) or 36–72–72 triangles (like DAE, HBF and GEC).

This is easily seen if you remember that the exterior angles of the pentagon (as BAC) are all 108°, and the diagonals (as AD and AE) trisect the angle BAC, making the angles BAF, FAG, and GAC all 36°

Notice that when the short side of any of these triangles is transferred to a longer side and a point marked, the resulting two triangles are both Phi ratio triangles. For example, observe triangle AGC in Fig.7-9. AGC is a 36–36–108 triangle. When the distance AG is transferred with the compass to line AC at P, two triangles are formed: AGP, a 72–72–36 triangle, and PGC, a 36–36–108 triangle.

If the small pentagon in the center had its diagonals drawn they too would divide themselves in Phi ratio and so the process would continue. Smaller and smaller pentagons would be formed, each one rotating 36 degrees and having sides Φ times smaller than the previous one.

Where does this end? or begin? The answer is it doesn't!

The pentagon 'ratchets' downward into infinite smallness, and similarly ratchets upward, getting larger and larger. This can only occur because division of each line segment is absolutely perfect. Not almost perfect, but totally perfect. There is never any 'round-off' error.

Notice also that each large triangle (ADE for example) is a golden triangle.

The smaller triangles (AFG for example) are also golden mean triangles.

The pentagon itself can be considered to be a rotating golden mean triangle. If you look at triangle ADE you can see how it rotates 5 times around the center of the circle (O), each rotation being 72 degrees (5 * 72 = 360 for the full circle).

Imagine the pentagon rotating downwards, out of the page, becoming infinitely small. Envision it rotating upwards out of the page, becoming infinitely large. The vertices, as they rotate, form a spiral.

So the growth cycle of the pentagon is like this:

$$...,...\frac{1}{\Phi^7},\ \frac{1}{\Phi^6},\frac{1}{\Phi^5},\frac{1}{\Phi^4},\frac{1}{\Phi^3},\frac{1}{\Phi^2},\ \frac{1}{\Phi},1,\ \Phi,\Phi^2,\Phi^3,\Phi^4,\Phi^5,\Phi^6,\Phi^7,....,$$

Construction of the Pentagon, Part II

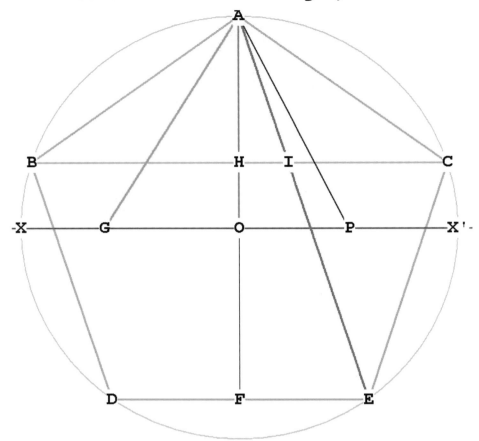

Fig. 7-6

Something very important happens during the construction of the pentagon: the triangle AOG.

AO = OX' = 1, which is the radius.

We bisected OX' at P, so OP = PX' = 1/2.

By the Pythagorean Theorem, $\overline{AP}^2 = \overline{AO}^2 + \overline{OP}^2 = 1 + \dfrac{1}{4} = \dfrac{5}{4}$.

AP = $\dfrac{\sqrt{5}}{2}$ = GP.

GX' = GP + PX', so GX' = $\dfrac{\sqrt{5}}{2} + \dfrac{1}{2}$.

This value is called Phi (Φ).

GO = GP - OP, so GO = $\dfrac{\sqrt{5}}{2} - \dfrac{1}{2}$.

This value is the inverse of Phi, or $\dfrac{1}{\Phi}$.

In Fig. 7-6, the distance AP (the hypotenuse of triangle AOP – See Note 1) is transferred to the line XX' at G, such that AP = PG. When this occurs, the line GOX' has been divided into Phi ratio at O.

Euclid called this the division into Mean and Extreme ratio.

The larger segment of GX' is OX', the smaller segment is GO.

$$GX' = \Phi = 1.618033989...$$

$$GO = \frac{1}{\Phi} = 0.618033989...$$

The triangle AOG has it's larger side (OA)= 1 and its smaller side (OG) = $\frac{1}{\Phi}$. Triangle AOG is a

Phi right triangle (as explained in Appendix C).

The hypotenuse of triangle AOG, AG, is the result of the division of GOA into the Phi ratio.

What is AG? By the Pythagorean theorem,

$$AG^2 = OA^2 + OG^2 = 1 + \frac{1}{\Phi^2} .$$

$$AG = \sqrt{1 + \frac{1}{\Phi^2}} .$$

AG = AB, the side of the pentagon. This is the secret to the design of the pentagon -- its sides are a result of a Phi ratio triangle. In the previous section we drew diagonals within the pentagon, and discovered that every single one of them intersects the others in phi ratio, forming smaller and smaller pentagons within each other, and forming many phi ratio triangles. The intersection of diagonals AE and BC, for example, divide each other in Phi ratio at I, and form the Phi ratio triangle AIC. ABC is also a phi ratio triangle, as is triangle ACE!

What is the relationship between the radius of the circle enclosing the pentagon, and the side of the pentagon?

We know that AB = side of pentagon(s) = $\sqrt{1 + \frac{1}{\Phi^2}}$ x radius.

OA = OB = OC = OD = OE = radius.

so $r = \dfrac{1}{\sqrt{1 + \dfrac{1}{\Phi^2}}} \times s$.

$1 + \dfrac{1}{\Phi^2} = \dfrac{\Phi^2 + 1}{\Phi^2}$. So we rewrite $r = \dfrac{\Phi}{\sqrt{\Phi^2 + 1}} \times s = 0.850650808s.$

What is the height of the pentagon AF?

$$AF^2 = AE^2 - FE^2 = \Phi^2 s^2 - \frac{1}{4} s^2 = \frac{4\Phi^2 - 1}{4} * s^2 = \frac{\Phi^2(\Phi^2 + 1)}{4} s^2$$

So h = AF = square root of that = $\dfrac{\Phi\sqrt{\Phi^2 + 1}}{2} \times s = 1.538841769.$

What is AH? AH bisects BC, a diagonal of the pentagon. We know that the diagonal of a pentagon is Φ * side of the pentagon. Therefore

$$HC = \frac{1}{2} \times \Phi \times s = \frac{\Phi}{2}s.$$

$$\overline{AH}^2 = \overline{AC}^2 - \overline{HC}^2 = s^2 - \left(\frac{\Phi}{2}s\right)^2 = \frac{4 - \Phi^2}{4}s^2$$

$$AH = \frac{\sqrt{4 - \Phi^2}}{2}s = \frac{\sqrt{\Phi^2 + 1}}{2\Phi}s = 0.587785252s.$$

What is OF?

$$OF = AF - AO = \frac{\Phi\sqrt{\Phi^2 + 1}}{2}s - \frac{\Phi}{\sqrt{\Phi^2 + 1}}s = \frac{\Phi(\Phi^2 + 1) - 2\Phi}{2\sqrt{\Phi^2 + 1}}s = \frac{\Phi(\Phi + 2) - 2\Phi}{2\sqrt{\Phi^2 + 1}}s =$$

$$= \frac{\Phi^2}{2\sqrt{\Phi^2 + 1}}s = 0.688190961s.$$

What is FH?

$$FH = AF - AH = \frac{\frac{\Phi\sqrt{\Phi^2 + 1}}{2}s - \frac{\sqrt{\Phi^2 + 1}}{2\Phi}s = \frac{\Phi^2\sqrt{\Phi^2 + 1} - \sqrt{\Phi^2 + 1}}{2\Phi} = \frac{\sqrt{\Phi^2 + 1}(\Phi^2 - 1)}{2\Phi}s}{= \frac{\Phi\sqrt{\Phi^2 + 1}}{2\Phi} = \frac{\sqrt{\Phi^2 + 1}}{2} = 0.951056517 \ s.}$$

$$\text{Note that : } FH \, / \, AH = \frac{\frac{\sqrt{\Phi^2 + 1}}{2}}{\frac{\sqrt{\Phi^2 + 1}}{2\Phi}} = \frac{\sqrt{\Phi^2 + 1}}{2} * \frac{2\Phi}{\sqrt{\Phi^2 + 1}} = \Phi!$$

Therefore, the diagonal BC divides the height of the pentagon, AF, in mean and extreme ratio.

What is the angle BAC?

We know that AC = 1, and that HC = $\frac{\Phi}{2}$, because BC = Φ. Therefore,

$$\sin(\angle HAC) = HC \, / \, CA = \frac{\Phi}{2}$$

$$\angle HCA = \arcsin\left(\frac{\Phi}{2}\right) = 54°$$

FA bisects $\angle BAC$, therefore $\angle BAC = 2 * \angle HCA = 108°$

Note 1:

Phi occurs geometrically within a 1, $\frac{1}{2}, \frac{\sqrt{5}}{2}$ triangle, when the short side distance is transferred

to the hypotenuse. That distance divides the long side of the triangle in Phi ratio, like so:

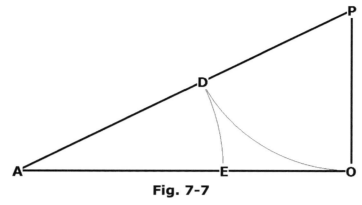

Fig. 7-7

OP is transferred to AP at D, and then AD is transferred to AO at E. AO has been divided in Mean and Extreme Ratio at E. The triangle AOP in Fig.7-7 is just such a triangle.

The Area of the Pentagon

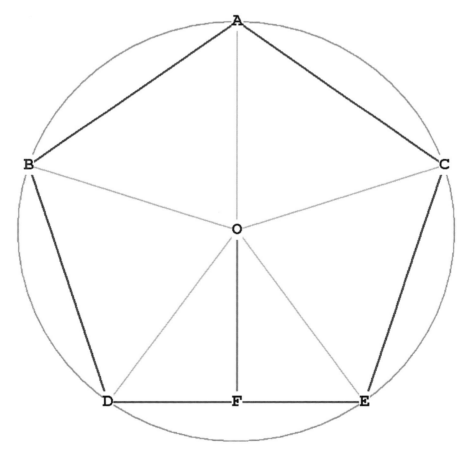

Fig. 7-8

The area of the pentagon is the combined area of the 5 triangles shown in Fig. 7-8, each of which has a vertex at O.

What is the area of each of these triangles?

Let's work with triangle ODE.

From   above we know that the height of triangle ODE, OF, is $\dfrac{\Phi^2}{2\sqrt{\Phi^2+1}}\,s$.

The area of each triangle = 1/2 * base * height:

$$Area_{1\,triangle} = 1/2 * s * \dfrac{\Phi^2}{2\sqrt{\Phi^2+1}}\,s = \dfrac{\Phi^2}{4\sqrt{\Phi^2+1}}\,s^2 .$$

Area of pentagon = 5 * area of each triangle.

$$Area_{pentagon} = \dfrac{5\Phi^2}{4\sqrt{\Phi^2+1}}\,s^2 = 1.720477401\ s^2 .$$

Part III - The Icosahedron and the Dodecahedron

Now that we understand the pentagon and Phi, we can begin to make sense of the Icosahedron and the Dodecahedron. Although the pentagonal nature of the dodecahedron is obvious, it turns out that the icosahedron is essentially pentagonal as well!

Chapter 8 – The Icosahedron

The icosahedron has 12 vertices, 20 faces and 30 sides. It is one of the most interesting and useful of all polyhedra. Buckminster Fuller based his designs of geodesic domes around the icosahedron.

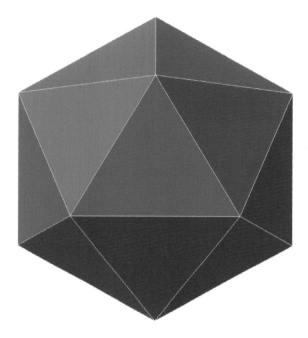

The icosahedron is built around the pentagon and the golden section. At first glance this claim may seem absurd, since every face of the icosahedron is an equilateral triangle. It turns out, however, that the triangular faces of the icosahedron result from its pentagonal nature.

First, we'll display 3 views of this polyhedron:

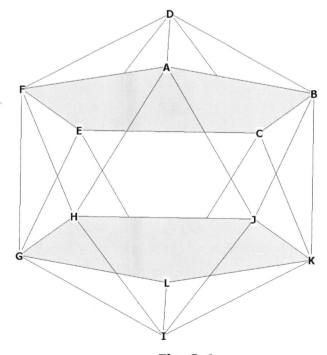

Fig. 8-1

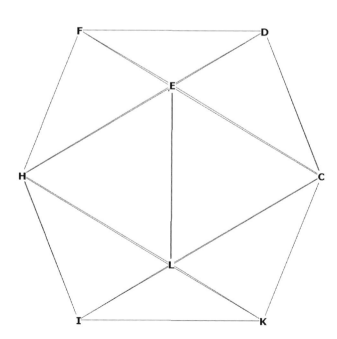

Fig. 8-2

Fig. 8-1 shows 2 of the internal pentagons of the icosahedron, LGFDC, and ABKIH.

Fig. 8-2 helps to show that the icosahedron is actually an interlocking series of pentagons. Notice the exterior pentagons at CDFHL, CE-HIK, EFHIL, and CDELK.

Fig. 8-3 shows a two dimensional 'shadow' of the icosahedron from the top down. You can see that the outer edges form a perfect deca-gon, formed by the two pentagons CDFGL and ABKIH.

In fact, every vertex of the icosahedron is the vertex of a pentagon.

OK, lets go through the usual analysis and then get on to the interesting stuff!

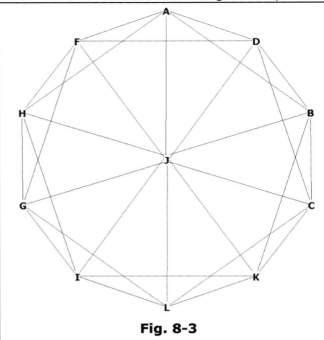

Fig. 8-3

First, let's calculate the volume of the icosahedron. Using the pyramid method we have 20 equilateral triangular faces which serve as the base to a pyramid whose topmost point is the centroid of the icosahedron. Fig. 84 shows some icosahedron faces with midpoints. All 20 faces will be connected to O to form 20 pyramids.

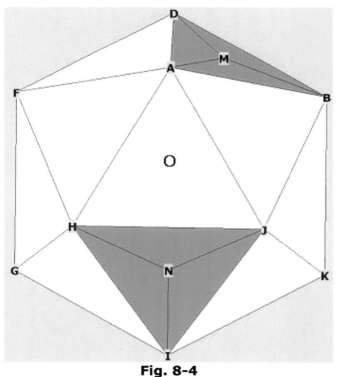

Fig. 8-4

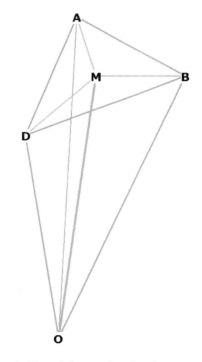

Fig. 8-5   1 icosahedral pyramid

The volume of each pyramid is 1/3 x (area of base) x (pyramid height).

The area of the base is the area of the equilateral triangle ADB.

The height of the pyramid is OM.

From *The Equilateral Triangle* we know this area is $\frac{\sqrt{3}}{4}s^2$.

All vertices of the icosahedron (as with all five of the regular solids) lie upon the surface of a sphere that encloses it. The radius of the circumsphere is O to any vertex, in this case,

r = OA = OB = OD = 1.

The height of the pyramid is h = OM.

In order to find h we need to recognize that any of the triangles OMA, OMB, OMD are right.

This is because OM is perpendicular to the plane of triangle ABD by construction.

AB = BD = AD = side of icosahedron = s.

Let's work with triangle OMD:

We know, from The Equilateral Triangle that DM = $\frac{1}{\sqrt{3}}$s.

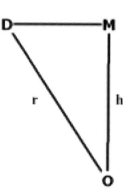

Fig. 8-6

In order to find h, we need to find OD = r, in terms of the side s of the icosahedron.

To do that, we have to recognize one of the basic geometric properties of the icosahedron.

Take a look at Fig. 8-1. DI, BG and FK are all diameters of the enclosing sphere around the icosahedron.

Notice rectangle BFGK and notice that BG, FK are both diagonals of it. Now notice that both FG and BK are both diagonals of the two pentagonal planes marked in gray.

We know from *Composition of the Pentagon* that the diagonal of a pentagon is Φ x side of pentagon.

Therefore FB = GK = Φ x side of icosahedron, since each side of the pentagon is a side of the icosahedron.   FB = Φ x s

FK is the diameter of the enclosing sphere around the icosahedron. OF = OK is the radius r, which we are trying to find.

FG = BK is the side s of the icosahedron.

$$\overrightarrow{FK}^2 = \overrightarrow{FB}^2 + \overrightarrow{BK}^2 = \Phi^2 s^2 + s^2 = (\Phi^2 + 1)s^2.$$

Fig. 8-7

FK = $s\sqrt{\Phi^2 + 1}$, r = d/2 and r = $\frac{\sqrt{\Phi^2 + 1}}{2}$s.

Now we can find h, the pyramid height.

Going back to Fig. 8-5 and Fig. 8-6, we can write:

$h^2 = OM^2 = r^2 - DM^2 =$

$$\frac{(\Phi^2 + 1)}{4} - \frac{1}{3} = \frac{3(\Phi^2 + 1) - 4}{12} = \frac{3(\Phi^2 - 1)}{12} = \frac{\Phi^4}{12} s^2.$$

$h = \dfrac{\Phi^2}{2\sqrt{3}} s$ is the height of the icosahedron pyramid.

$\text{Volume}_{1 \text{ pyramid}} = \dfrac{1}{3} \times \text{area of base} \times \text{height} = \dfrac{1}{3} \times \dfrac{\sqrt{3}}{4} s^2 \times \dfrac{\Phi^2}{2\sqrt{3}} s.$

So $V_{1 \text{ pyramid}} = \dfrac{\Phi^2}{24} s^3.$

There are 20 pyramids, 1 for each face so

$\text{Volume}_{\text{icosahedron}} = \dfrac{5\Phi^2}{6} s^3 = 2.181694991 s^3.$

What is the surface area of the icosahedron?

It is just the area of 1 face * 20 faces. The area of each face is, from above, $\dfrac{\sqrt{3}}{4} s^2$.

So $\text{Surface area}_{\text{icosahedron}} = 20 * \dfrac{\sqrt{3}}{4} s^2 = 5\sqrt{3}\ s^2 = 8.660254038\ s^2.$

We have already noted the relationship between the radius of the enclosing sphere and the side of the icosahedron:

$r = \dfrac{\sqrt{\Phi^2 + 1}}{2} s, s = \dfrac{2}{\sqrt{\Phi^2 + 1}} r.$ $r = .951056517s,\ s = 1.051462224r.$

The side or edge of the icosahedron is slightly larger than the radius.

What is the central angle of the icosahedron?

The central angle, $\angle DOB$, can be seen clearly from Fig. 8-5, and we diagram it in Fig. 8-8 at right.

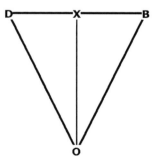

$OD = OB = r = \dfrac{\sqrt{\Phi^2 + 1}}{2} s.$ DB is the side of the icosahedron, s.

$\sin(\angle XOB) = \dfrac{XB}{OB} = \dfrac{\dfrac{1}{2}}{\dfrac{\sqrt{\Phi^2 + 1}}{2}} = \dfrac{1}{\sqrt{\Phi^2 + 1}}.$

$\angle XOB = \arcsin(\dfrac{1}{\sqrt{\Phi^2 + 1}}) = 31.7174744°.$ $\angle DOB = 2 * \angle XOB.$

Fig. 8-8

We recognize triangle OXB as our old friend the Phi Right Triangle. From this we know that OX / XB = Φ.

Central angle = 2 * $\angle XOB$ = 63.4349488°

Surface angles = 60°

What is the dihedral angle of the icosahedron?

The dihedral angle is the angle formed by the intersection of 2 planes:

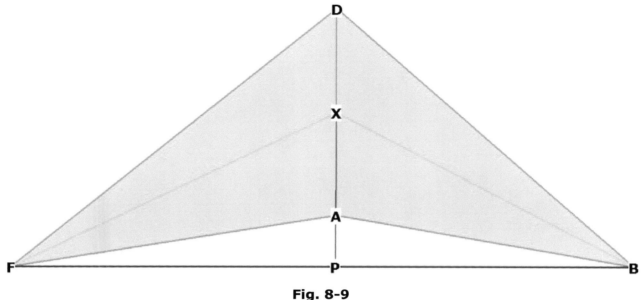

Fig. 8-9

The intersection of the 2 faces DFA and ABD forms the dihedral angle FXB. (See Fig.8-4 as well).

FX = XB we know, from The Equilateral Triangle to be $\frac{\sqrt{3}}{2}$ s.

FB = a diagonal of one of the pentagons. This can be seen in Fig. 8-1 as the diagonal of pentagon ABCEF.

Therefore FB = Φ * s and PB = $\frac{\Phi}{2}$ * s .

Triangles XPB and XPF are right by construction, so

$\angle$ PXB = 1/2 $\angle$ FXB.

$\sin(\angle PXB) = PB / XB = \dfrac{\frac{\Phi}{2}}{\frac{\sqrt{3}}{2}} = \dfrac{\Phi}{\sqrt{3}}$.

$\angle$ PXB = arcsin($\frac{\Phi}{\sqrt{3}}$) = 69.09484258°

Dihedral angle = 138.1896852°

Now let's figure out the distances from the centroid of the icosahedron to any vertex, to any mid-face, and to any mid-side.

We already have the first two (see Figure 10). OD = OA = OB = r = $\dfrac{\sqrt{\Phi^2 + 1}}{2}$ s

OM = h = $\dfrac{\Phi^2}{2\sqrt{3}}$ s.

Now we need to find, for example, OX.

If we lay a 3D model of the icosahedron on one of its sides, we can see that a line through the centroid O is perpendicular to that side. So the triangle OXB is right.

We know AB = s, so XB = (1/2)s. OB = r, so

$$OX^2 = OB^2 - BX^2 \quad = \frac{\Phi^2+1}{4}s^2 - \frac{1}{4}s^2 = \frac{\Phi^2+1-1}{4}s^2 = \frac{\Phi^2}{4}s^2.$$

$$OX = \frac{\Phi}{2}s.$$

distance from centroid to mid-face = $\dfrac{\Phi^2}{2\sqrt{3}}s$ = 0.755761314s.

distance from centroid to mid-side = $\dfrac{\Phi}{2}s$ = 0.809016995s.

distance from centroid to a vertex = $\dfrac{\sqrt{\Phi^2+1}}{2}s$ =

0.951056517s.

Now let's get to the interesting stuff!

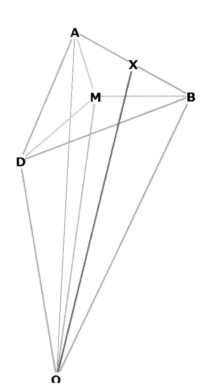

Fig. 8-10

Going back to Figure 1, we can see that the icosahedron is composed of interlocking pentagonal 'caps.'

Look at D-ABCEF and I-GHJKL to see this more clearly. Of course , EVERY vertex of the icosahedron is the top of a pentagonal cap, not just D and I.

Let's analyze this cap:

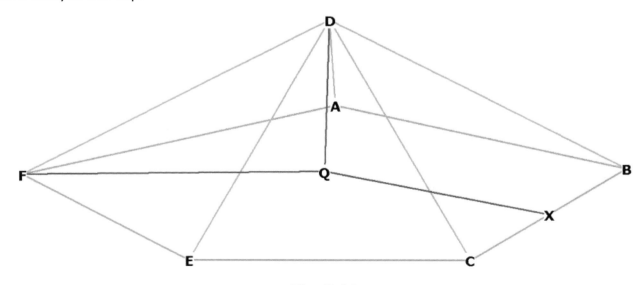

Fig. 8-11

Triangle DQF is right, by construction. D is directly over the point Q in the icosahedron.

How far off the plane of pentagon AFECB is D?

In other words, what is the distance DQ?

First, realize there is a circle around pentagon AFECB, even though I haven't draw it here. QF is the radius of that circle. We know from   Construction of the Pentagon, Part 2 that this radius

$$r = FQ = \frac{\Phi}{\sqrt{\Phi^2 + 1}}s.$$ Here, s = the side of the icosahedron.

DF is a side of the icosahedron, so DF = s. Therefore we write

$$DQ^2 = DF^2 - FQ^2 = s^2 - \frac{\Phi^2}{\Phi^2 + 1}s^2 = \frac{(\Phi^2 + 1) - \Phi^2}{\Phi^2 + 1}s^2 = \frac{1}{\Phi^2 + 1}s^2.$$

$$DQ = \frac{1}{\sqrt{\Phi^2 + 1}}s.$$

Hmmm, this is looking interesting. Let's compare FQ to DQ.

$$FQ / DQ = \frac{\dfrac{\Phi}{\sqrt{\Phi^2 + 1}}}{\dfrac{1}{\sqrt{\Phi^2 + 1}}} = \Phi!$$

In order to form equilateral triangles with D and the other vertices of the pentagon, D has to be raised off the pentagonal plane AFECB by the division of the radius of the pentagon (FQ) in Mean and Extreme Ratio.

The triangle DQF is therefore a Phi based triangle, specifically, a $1, \Phi, \sqrt{\Phi^2 + 1}$ triangle. From The Phi Right Triangle we know that $\angle DFQ = 31.71747441°$.

Is this surprising? It was to me! I didn't expect something that was composed entirely of equilateral triangles to have any relationship to Φ. An equilateral triangle is $\sqrt{3}$ geometry, Φ is $\sqrt{5}$ geometry.

Here we have Fig. 8-1 basically, with the mid-face points of the two internal pentagons marked off as Q and Z.

We have already seen that DQ is

$\dfrac{1}{\sqrt{\Phi^2+1}}$ with respect to the side of the icosa-

hedron.

That means $IZ = \dfrac{1}{\sqrt{\Phi^2+1}}$ s as well, since IZ

= DQ.

What about OQ = OZ? and QZ? How do all of these distances relate to the diameter of the enclosing sphere, DI? Remember that the distance DI = FK = BG, etc., is the diagonal of any of the Φ rectangles of which the ico-sahedron is composed.

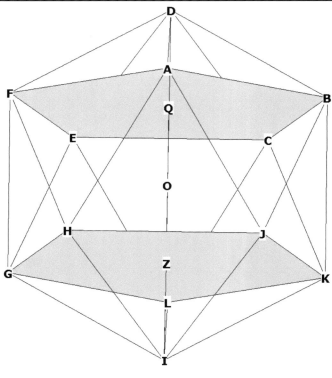

Fig. 8-13

One of these Φ rectangles is clearly visible in Fig. 8-13 as BFGK. We know this is a Φ rectangle because FB and GK are the diagonal of the pentagon ABCEF, which sides are the sides of the icosahedron.

In fact, if you place 3 of these rectangles perpendicular to each other, the 12 corners of the 3 rectangles are the vertices of the icosahedron!   FB = Φ

I have copied Fig. 8-7 from above.

FK = diameter, OF = radius of enclosing sphere.

We know from above that $\underline{d = FK} = DI = \sqrt{\Phi^2+1}$ s.

We have already figured out DQ = ZI. So QZ = DI - 2*DQ.

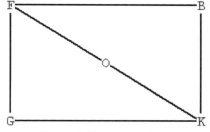

Fig. 8-7, repeated

$$QZ = \sqrt{\Phi^2+1} - \left(2*\frac{1}{\sqrt{\Phi^2+1}}\right) = \frac{\Phi^2+1-2}{\sqrt{\Phi^2+1}} = \frac{(\Phi^2-1)}{\sqrt{\Phi^2+1}} = \frac{\Phi}{\sqrt{\Phi^2+1}} \text{ s.}$$

Now,

$$QZ / DQ = \frac{\dfrac{\Phi}{\sqrt{\Phi^2+1}}}{\dfrac{1}{\sqrt{\Phi^2+1}}} = \Phi!$$

So DZ is divided in EMR at Q (See Figure 13).

One interesting fact appears here: FQ = QZ. This means that the distance from one pentagonal plane to the other is precisely the radius of the circle that encloses the pentagon ABCEF.

What is OQ?

It looks like OQ is just one-half that of QZ, or $\dfrac{\Phi}{2\sqrt{\Phi^2+1}}$ s. But is it? Let's find out.

We know OD = r = $\dfrac{\sqrt{\Phi^2+1}}{2}$ from above.

OQ = OD - DQ = $\dfrac{\sqrt{\Phi^2+1}}{2} - \dfrac{1}{\sqrt{\Phi^2+1}} = \dfrac{\Phi^2+1-2}{2\sqrt{\Phi^2+1}} =$
$\dfrac{(\Phi^2+1)}{2\sqrt{\Phi^2+1}} = \dfrac{\Phi}{2\sqrt{\Phi^2+1}}$ s.

OQ = $\dfrac{\Phi}{2\sqrt{\Phi^2+1}}$ s. We could have gotten this more easily from the fact that DQ = $\dfrac{1}{\sqrt{\Phi^2+1}}$ s.

Yes, OQ is exactly one-half QZ.

Also, OQ / DQ $= \dfrac{\Phi}{2}$ S = 0.951056517

Let DQ = 1

Table of Relationships

D_____		D_____	
1			
Q_____	Q_____		Q_____
$\Phi/2$		$\Phi+1 = \Phi^2$	
O_____	Φ		
$\Phi/2$			$\Phi+1 = \Phi^2$
Z_____	Z_____	Z_____	
1			
I_____			I_____

Fig.8-14. Central axis (diameter) relationships   diameter = DI

So DZ is divided in EMR by Q, IQ is divided in EMR by Z.

All of these relationships come from the pentagon!

On the outside of the icosahedron, we see equilateral triangles. But the guts of this polyhedron come from pentagonal relationships. The equilateral triangles come about from the lifting of the pentagonal 'cap' off the pentagonal plane.

There is now no question that the basis for the construction of the icosahedron is the pentagon. Or is there?

Take a look at this view of the icosahedron in Fig. 8-15. As you can see, the 2 pentagonal planes in the middle have magically disappeared and become equilateral triangles. All we have done is placed the icosahedron on one of its faces.

It seems that all of our work is wrong, except

- The sides of the equilateral triangles EHB and JGC are all diagonals of pentagons!
- EH is a diagonal of pentagon FEJIH, EB is a diagonal of pentagon DEJKB, and HB is a diagonal of pentagon AHIKB.
- The equilateral triangle faces of the icosahedron are a by-product of how the pentagons interlock. This is shown in Fig. 8-16

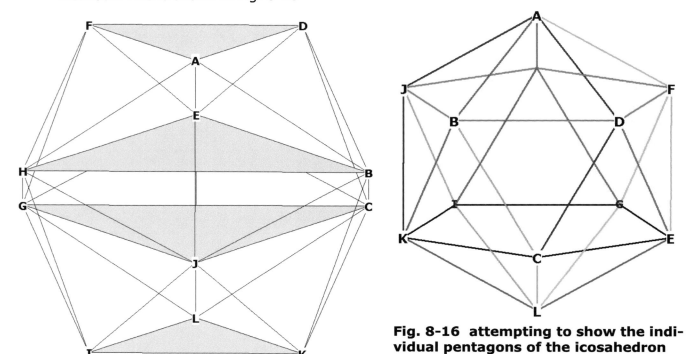

Fig. 8-15

Fig. 8-16 attempting to show the individual pentagons of the icosahedron

Because the pentagons are interlocking, there is duplication of colors, which makes the individual pentagons a bit difficult to identify.

Begin with the two pentagons BDFHJ, in green, and CEGIK, in black.

Insert pentagon AJKCD, 4 of which sides are in blue.

Insert pentagon ABKIH, 4 of which sides are marked in red.

Insert pentagon BCLIJ, 4 of which sides are marked in orange.

Insert pentagon ADEGH, 2 of which sides are marked in magenta.

Insert pentagon AFECB, 2 of which sides are marked in gold.

Insert pentagon CDFGL, 2 of which sides are marked in cyan (sky blue)

The side KL in white is part of the pentagon KLGHJ.

The side LE is part of the pentagon ELIHF.

Icosahedron Reference Charts

Volume (edge)	Volume in Unit Sphere	Surface Area (edge)	Surface Area in Unit Sphere
$2.181694991s^3$	$2.53615071 r^3$	$8.660254038 s^2$	$9.574541379 r^2$

Central Angle:	Dihedral Angle:	Surface Angle:
63.4349488°	138.1896852°	60°

Centroid To: Vertex	Centroid To: Mid–Edge	Centroid To: Mid–Face
1.0 r	0.850650809 r	0.794654473 r
0.951056517 s	0.809016995 s	0.755761314 s

Side / radius
1.051462224

Chapter 9 – The Dodecahedron

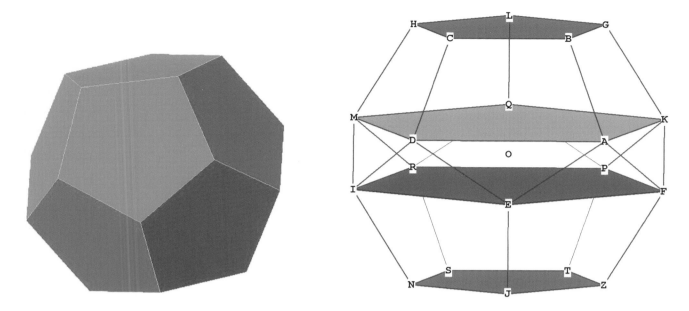

Fig. 9-1

The dodecahedron has 30 edges, 20 vertices and 12 faces. Dodeca is a prefix meaning "twelve." The dodecahedron is composed entirely of pentagons.

The dodecahedron is pentagonal both inside and out, as can be seen from Fig. 9-1. Like the icosahedron, it has many golden section relationships, which we shall see.

The dodecahedron is even more versatile then the icosahedron. The icosahedron contains $\sqrt{3}$ and $\sqrt{5}$ geometry, but the dodecahedron contains $\sqrt{2}$, $\sqrt{3}$ and $\sqrt{5}$ geometry!

This view of the dodecahedron is significant in that it shows the 2 dimensional shadow of the decagon. The decagon itself is based upon the pentagon, the building block of the dodecahedron.

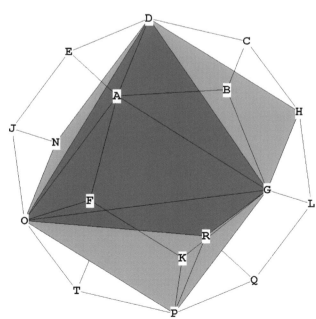

Fig. 9-2 Cube and tetrahedron in dodecahedron. Cube in gray, tetrahedron in green

Figs 9-2 and 2A show how a cube and a tetrahedron can be placed inside a do-decahedron, directly upon its vertices. The cube, octahedron and tetrahedron are all based on root 2 and root 3 ge-ometry: The relationship of the side of the cube to the radius of its enclosing sphere is $r = \dfrac{2}{\sqrt{3}}$ s. For the tetrahedron,

$r = \dfrac{2\sqrt{2}}{\sqrt{3}}$ s. For the octahedron, $r = \dfrac{1}{\sqrt{2}}$ s

The dodecahedron is capable of elegantly sustaining these $\sqrt{2}$ and $\sqrt{3}$ relation-ships, along with its own many $\sqrt{5}$ relationships

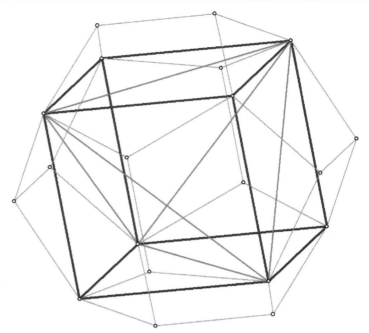

Fig. 9-2A: **Another view of tetrahedron (green) inside cube (blue) inside dodecahedron (orange)**

Notice that the octahedron fits precisely on the bisected sides of the tetrahedron. The icosahedron cannot contain any of the other 5 solids 'nicely' on its vertices. The icosahedron and the dodecahedron are duals (as are the cube and the octa-hedron). By 'dual' is meant that if you put a vertex in the middle of every face and connect the lines, you get the dual. By placing a vertex at the middle of each face of the dodecahedron you get an icosahedron, and vice-versa.

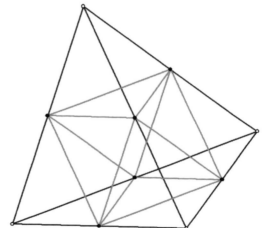

Fig.9-3: octahedron inside tetrahedron

Fig. 9-4 shows the dual nature of the icosahedron and dodecahedron.

To create the dodecahedron, all we did was draw lines from each vertex of the icosahedron to every other vertex. The vertices of the do-decahedron are at the intersection points. We could just as easily have found the vertices of the dodecahedron by drawing lines on every triangular face of the icosahedron. Where those lines intersect is the center of the face, and a vertex of the dodecahedron. That occurs because the dodecahedron has 12 faces and the icosahedron has 12 vertices. I've shown the dual this way because the diagram is less cluttered.

Now for the standard analysis:

What is the volume of the dodecahedron?

We will use the pyramid method.

Fig. 9-4 Duals: dodecahedron inside ico-sahedron

There are 12 pentagonal pyramids, 1 for each face, each pyramid beginning at O, the centroid. See Fig. 9-1 and Fig. 9-5.

The volume of any n-sided pyramid is 1/3 * area of base * pyramid height.

First we need to get the area of the base, which is the area of each pentagonal face:

The area of the pentagon is the area of the 5 triangles which compose it.

From Area of Pentagon we know

$$a = \frac{5\Phi^2}{4\sqrt{\Phi^2 + 1}} s^2 = 1.720477401 \, s^2.$$

Now we need to find the height of the pyramid, OU. To do that, we need to find the distance from O to a vertex, lets say, OH. This distance will be the hypotenuse of the right triangle OUH. Since we already know UH, we can then get OU from the Pythagorean Theorem.

Imagine a sphere surrounding the dodecahe-dron and touching all of its vertices. OH is just the radius of the enclosing sphere. If you look at Fig. 9-6, HOZ = GON = diameter. There is a line through HOZ to show the diameter.

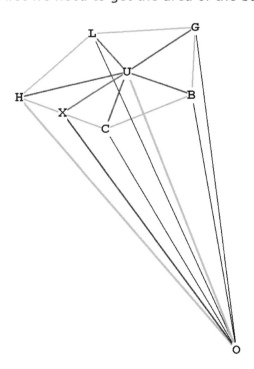

Fig. 9- 5   One pyramid on face BCHLG

Now look at the rectangle MIFK. The diagonal of it, MF, is also a diameter (MF = HZ). Notice that the long sides of the rectangle, MK, and IF, are diagonals of the two large pentagons ADMQK and FEIRP.

We know from Composition of the Pentagon that the diagonal of a pentagon is Φ x side of pentagon.

We also can see from Fig. 9-6 and more clearly in Fig. 9-1 that the sides of the large pentagons themselves are diagonals of the pentagonal faces of the dodecahedron! (For instance, DA is a diagonal of the face ABCDE).

That means each side of the large pentagons is

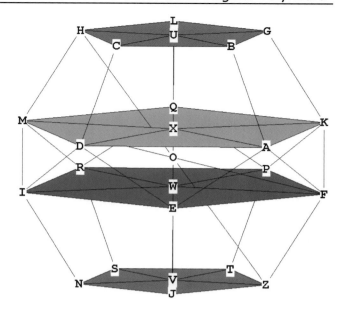

Fig. 9-6

Φ x s and that MK (or any of the diagonals of a large pentagon) is Φ x Φ x s. So MK = Φ^2s. In fact, like the icosahedron, the dodecahedron is composed of rectangles divided in Extreme and Mean Ratio. In the icosahedron, we found these rectangles to be Φ rectangles.

In the dodecahedron, they are Φ^2 rectangles.

In rectangle MKIF, MK = IF = Φ^2s , MI = KF = s , as shown in Fig. 9-7.

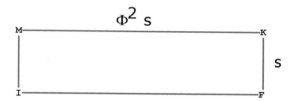

Fig. 9-7: showing vertices of the Φ^2 rectangle MKIF.

There are 30 sides to the dodec, and therefore 15 different Φ^2 rectangles.

All of this as explanation of finding the distance OH from Fig. 9-5! Because we are not using trigonometry, we need OH in order to get the pyramid height, OU in Fig.9-5. Notice that MHFZ is also a Φ^2 rectangle and that HZ is the diagonal of it. If we can find HZ, then OH is just 1/2 of that.

d² = HZ² = MZ² + HM² = $\Phi^4 + 1 = (3\Phi + 2) + 1 = 3\Phi + 3 = 3(\Phi + 1) = 3\Phi^2$.

diameter = HZ = $\sqrt{3}\Phi \times$ s.

$$OH = r = \frac{\sqrt{3\Phi}}{2} s, \quad s = \frac{2}{\sqrt{3\Phi}} r.$$

Now we can find OU, the height of the pyramid.

From *Construction of the Pentagon, Part II* we know the distance mid-face to any vertex of a pentagon $= \dfrac{\Phi}{\sqrt{\Phi^2 + 1}} * s.$

So UH $= \dfrac{\Phi}{\sqrt{\Phi^2 + 1}}.$

$$h^2 = \overline{OU}^2 = \overline{OH}^2 - \overline{UH}^2 = \begin{aligned} &= \dfrac{3\Phi^2}{4}s^2 - \dfrac{\Phi^2}{\Phi^2 + 1} = \dfrac{3\Phi^4 + 3\Phi^2 - 4\Phi^2}{4(\Phi^2 + 1)} \\ &= \dfrac{3\Phi^4 - \Phi^2}{4(\Phi^2 + 1)} = \dfrac{\Phi^2(3\Phi^2 - 1)}{4(\Phi^2 + 1)} \\ &= \dfrac{\Phi^4 * \Phi^2}{4(\Phi^2 + 1)} = \dfrac{\Phi^6}{4(\Phi^2 + 1)}. \end{aligned}$$

$$h = OU = \dfrac{\Phi^3}{2\sqrt{\Phi^2 + 1}}\, s.$$

The volume of 1 pyramid = 1/3 * area base * pyramid height =

$$1/3 * \dfrac{5\Phi^2}{4\sqrt{\Phi^2 + 1}} s^2 * \dfrac{\Phi^3}{2\sqrt{\Phi^2 + 1}} s = \dfrac{5\Phi^5}{24(\Phi^2 + 1)} s^3.$$

$$\text{Volume}_{\text{dodeca}} = 12 * \text{Volume}_{\text{1pyramid}} = \dfrac{5\Phi^5}{2(\Phi^2 + 1)} s^3 = 7.663118963 s^3. \ \text{Or}$$

$$\text{Volume}_{\text{dodeca}} = \dfrac{5\Phi^5}{2(\Phi^2 + 1)} s^3 = \dfrac{5\Phi^5}{2(\Phi^2 + 1)} * \left(\dfrac{2}{\sqrt{3}\Phi} r\right)^3 = \dfrac{20\Phi^2}{3\sqrt{3}(\Phi^2 + 1)} = 2.78516386 r^3.$$

Note that (from Fig. 9-5) OU / UX $= \Phi$.

What is the surface area of the dodecahedron? It is

12 faces * area of face $= 12s * \dfrac{5\Phi^2}{4\sqrt{\Phi^2 + 1}} s = \dfrac{15\Phi^2}{\sqrt{\Phi^2 + 1}} s^2 = 20.64572881\, s^2$, or

$$\dfrac{15\Phi^2}{\sqrt{\Phi^2 + 1}} * \left(\dfrac{2}{\sqrt{3}\Phi} r\right)^2 = \dfrac{20}{\sqrt{(\Phi^2 + 1)}} r^2 = 10.51462224\, r^2$$

What is the central angle of the dodecahedron? From Fig. 9-5, this is (for example)
$\measuredangle$ HOC:

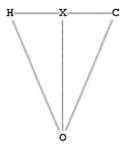

Fig. 9-8: dodecahedron central angle

HC = side of dodecahedron, so XH = (1/2)s.

OH = radius = one half HZ = $\dfrac{\sqrt{3\Phi}}{2}$ s .

$\sin(\angle XOH) = XH\,/\,OH\ =\ \dfrac{\frac{1}{2}}{\frac{\sqrt{3\Phi}}{2}} = \dfrac{1}{\sqrt{3\Phi}}.$

$\angle XOH = \arcsin(\dfrac{1}{\sqrt{3\Phi}}) = 20.90515744°$

$\angle HOC$ = central angle = $41.81031488°$

Since each face of the dodecahedron is a pentagon, the surface angle = 108°

While we're at it, lets get OX, the distance from the centroid to any mid- edge.

$$\overline{OX}^2 = \overline{OH}^2 - \overline{HX}^2 = \dfrac{3\Phi^2}{4}s^2 - \dfrac{1}{4}s^2 = \dfrac{3\Phi^2-1}{4}s^2 = \dfrac{\Phi^4}{4}s^2.$$

$OX\ =\ \dfrac{\Phi^2}{2}s = 1.309016995s$

What is the dihedral angle of the dodecahedron?

The dihedral angle is $\angle AXH$. AH is one of the

long sides of any of the 15 Φ^2 rectangles

which compose the dodec. AX and HX are the

height h of the pentagon.

We know from   Construction of the Pentagon

Part 2 that the height h of the pentagon

is: $\dfrac{\Phi\sqrt{\Phi^2+1}}{2}$ s.

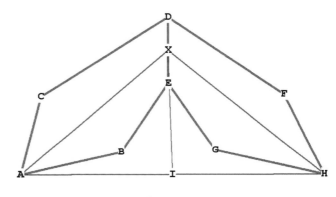

Fig. 9-9

We know from Figure 7 that AH is $\Phi^2 s$, so IH = $\dfrac{\Phi^2}{2}$ S.

$\sin(\angle IXH)\ = IH\,/\,XH\ =\ \dfrac{\frac{\Phi^2}{2}}{\frac{\Phi\sqrt{\Phi^2+1}}{2}} = \dfrac{\Phi}{\sqrt{\Phi^2+1}}.$

We recognize this ratio as our friend the Phi Right Triangle with sides in ratio of 1, Φ, $\sqrt{\Phi^2+1}$.

$\angle IHX\ =\ \arcsin(\dfrac{\Phi}{\sqrt{\Phi^2+1}}) = 58.28252558°.$

Dihedral angle AXH = 2 * $\angle IXH$,

So dihedral angle AXH = $116.5650512°$.

Let's compare distances:

Distance from centroid to mid-face (h) $= \dfrac{\Phi^3}{2\sqrt{\Phi^2+1}}s = 1.113516365s.$

Distance from centroid to mid-edge $= \dfrac{\Phi^2}{2} s = 1.309016995$.

Distance from centroid to vertex $= \dfrac{\sqrt{3}\Phi}{2} s = 1.401258539$.

Go back to Fig. 9-6. We have colored the 4 internal pentagonal planes of the dodecahedron. U,X, W and V are the centers of these 4 planes which line up with the centroid O.
What is the distance UX = WV? What is XW?
If we can find these out we can figure out more deeply how the dodecahedron is constructed.

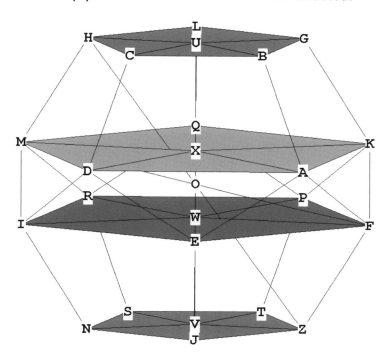

In Fig. 9-6 we can see that UH on the top plane is the distance from the pentagon center to a vertex.

On plane ADMQK, XM is parallel to UH and is also the distance from that pentagon center to a vertex. UH is connected to XM by HM, a side of the pentagon face CDIMH.

So we have a quadrilateral UHMX, with UH parallel to XM. From here we can derive UX, the distance between the two planes

Fig.9-6, repeated

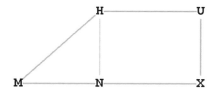

Fig. 9-10: dodecahedron planar distance UX.

We know from Construction of the Pentagon Part 2 that the distance from the center of pentagon to a vertex $= \dfrac{\Phi}{\sqrt{\Phi^2 + 1}} s$.

Therefore UH $= \dfrac{\Phi}{\sqrt{\Phi^2 + 1}} s$.

The side of the large pentagon ADMQK in Figure 6 is, as we have seen, a diagonal of a dodecahedron face and so the side of the large pentagon = $\Phi*s$. Therefore, $XM = \dfrac{\Phi^2}{\sqrt{\Phi^2+1}}s$.

Therefore, $XM = \Phi*UH$, and XM is divided in Mean and Extreme Ratio at N (See Figure 10).

$MN = XM - XN = \dfrac{\Phi^2}{\sqrt{\Phi^2+1}} - \dfrac{\Phi}{\sqrt{\Phi^2+1}} = \dfrac{1}{\sqrt{\Phi^2+1}}.$

HM = s. Triangle MNH is right by construction.

So $\overline{HN}^2 = \overline{HM}^2 - \overline{MN}^2 = s^2 - \dfrac{1}{\Phi^2+1}s^2 = \dfrac{\Phi^2+1-1}{\Phi^2+1}s^2 = \dfrac{\Phi^2}{\Phi^2+1}s^2.$

$HN = UX = \dfrac{\Phi}{\sqrt{\Phi^2+1}}s$

Notice: UH = UX. So the dodecahedron is designed such that the distance to the 2 large pentagonal planes from the top or bottom faces is exactly equal to the distance between the center and a vertex of any of the faces of the dodecahedron.

This relationship is precisely what we saw in the icosahedron! That makes sense because the two are duals of each other.

The difference is that the dodecahedron is entirely pentagonal, both internally, and externally, on its faces.

What is the distance XW between the 2 large pentagonal planes ADMQK and FEIRP?

Figure 6 is misleading, it looks like the distance must be MI or KF, the dodecahedron side, but it isn't.

We already have enough information to establish this distance.

$UX = VW$. $UV = 2*\text{height of any pyramid} = \dfrac{2\Phi^3}{2\sqrt{\Phi^2+1}} = \dfrac{\Phi^3}{\sqrt{\Phi^2+1}}.$

So $XW = UV - 2*UX = \dfrac{\Phi^3}{\sqrt{\Phi^2+1}} - \dfrac{2\Phi}{\sqrt{\Phi^2+1}} = \dfrac{\Phi^3-2\Phi}{\sqrt{\Phi^2+1}}.$

$XW = \dfrac{1}{\sqrt{\Phi^2+1}}s.$

Notice $UX / XW = \Phi$.

UW is divided in Mean and Extreme Ratio at X.

$XV / WV = \Phi$. XV is divided in Mean and Extreme Ratio at W.

Let's make a chart of these planar distances along the diameter of the dodecahedron as we did with the icosahedron: (see Figure 6):

Relative Chart of Distances

Pentagonal Planes of Dodecahedron Relative to the side of the dodecahedron

U_____   U_____   U_____

$$\frac{\Phi}{\sqrt{\Phi^2+1}}=0.85065$$

$$\frac{\Phi^3}{2\sqrt{\Phi^2+1}}=1.11352$$

X_____   X_____

$$\frac{\Phi^2}{\sqrt{\Phi^2+1}}=1.3764$$

$$\frac{1}{2\sqrt{\Phi^2+1}}=0.2629$$

O_____

$$\frac{1}{\sqrt{\Phi^2+1}}=0.5257$$

O_____

$$\frac{1}{2\sqrt{\Phi^2+1}}=0.2629$$

W_____   W_____   W_____

$$\frac{\Phi^3}{2\sqrt{\Phi^2+1}}=1.11352$$

$$\frac{\Phi}{\sqrt{\Phi^2+1}}=0.85065$$

V_____   V_____

Here is a table of these relationships, letting UX = 1:

U_____	**U_____**	**U_____**	
1			
			$\frac{\Phi^2}{2}$
X_____	X_____	Φ	
$\frac{1}{2\Phi}$			
$\frac{1}{2\Phi}$			
O_____	$\frac{1}{\Phi}$		O_____
$\frac{1}{2\Phi}$			
W_____	W_____	W_____	
V_____			

What is the distance from U to the top of the sphere, and from V to the bottom of the sphere?

The diameter of the enclosing sphere is HZ, or $\sqrt{3}\Phi * s$.

Let T' be the top of the sphere and B' be the bottom of the sphere. Refer to Fig. 9-11 and Fig. 9-6.

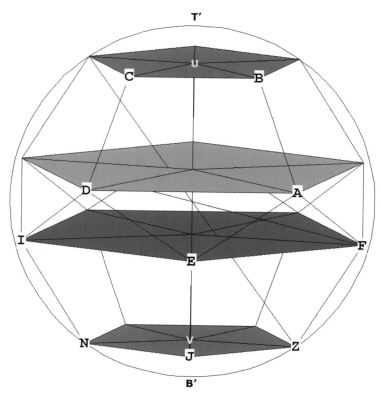

Fig. 9-11

If the radius is $\dfrac{\sqrt{3}\Phi}{2}$ and the distance OU is $\dfrac{\Phi^3}{2\sqrt{\Phi^2+1}}$, then

$$UT' = VB' = \frac{\sqrt{3}\Phi}{2} - \frac{\Phi^3}{2\sqrt{\Phi^2+1}} = 0.287742174s.$$

Finally, let us demonstrate how the dodecahedron may be constructed from the interlocking vertices of 5 tetrahedrons. We have already seen how the cube fits inside the dodecahedron, and how 2 interlocking tetrahedron may be formed from the diagonals of the cube. As Buckminster Fuller has pointed out, however, the cube and the dodecahedron are structurally unsound unless bolstered by the additional struts supplied by the tetrahedron. Fuller concludes logically that the tetrahedron is the basic building block of Universe; yet it is the dodecahedron that provides the blueprint and forms the structure for the interlocking tetrahedrons. The dodecahedron unites the geometry of crystals and lattices (root 2 and root 3) with the geometry of Phi (root 5), found in the biology of organic life.

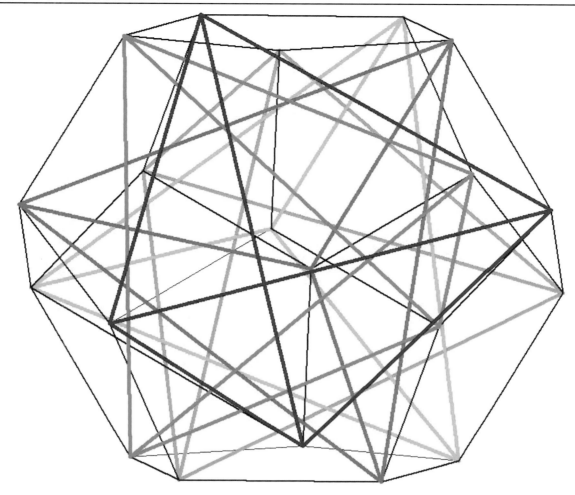

Conclusions:

The dodecahedron is entirely pentagonal, consisting of the $\sqrt{5}$ geometry of Phi. Yet it contains the $\sqrt{3}$ and $\sqrt{2}$ geometry of the cube, tetrahedron, and octahedron.

Remarkably, the sides of the cube are Φ x side of the dodecahedron, because the cube side is the diagonal of a pentagonal face. Here is the key to the relationship of the first three Regular Solids and the much more complex icosahedron and dodecahedron.

Later on in this book we will discover a remarkable polyhedron that defines the relationship and provides the proper nesting for all 5 Platonic Solids directly on its vertices. If a polyhedron could be called exciting, this one is IT! If you can't wait, go to the last chapter of the book.

Dodecahedron Reference Tables

Volume (edge)	Volume in Unit Sphere	Surface Area (edge)	Surface Area in Unit Sphere
7.663118963 s³	2.785163863 r³	20.64572881 s²	10.51462224 r²

Central Angle:	Dihedral Angle:	Surface Angle:
41.81031488°	116.5650512°	108°

Centroid To: Vertex	Centroid To: Mid–Edge	Centroid To: Mid–Face
1.0 r	0.934172359 r	0.794654473 r
1.401258539 s	1.309016995 s	1.113516365 s

Side / radius
0.713644179

Part IV – Semi-Regular Polyhedra

- The Cube Octahedron
- The Star Tetrahedron
- The Rhombic Dodecahedron
- The Icosa-Dodecahedron
- The Rhombic Triacontahedron

Chapter 10 – The Cube Octahedron

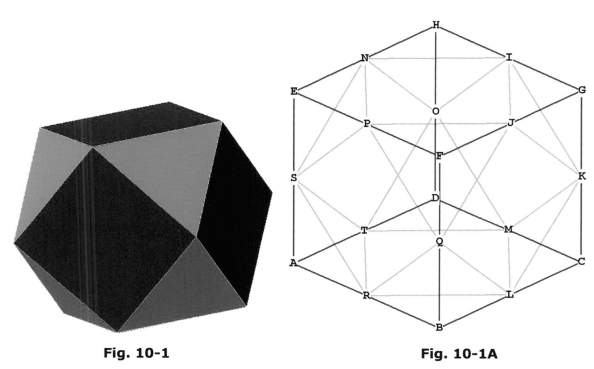

| Fig. 10-1 | Fig. 10-1A |

Fig. 10-1A shows the cube octahedron (yellow) inside the cube (blue). The cube octahedron is sometimes called the *vector equilibrium*. That is what Buckminster Fuller called it.   It has 12 vertices, 14 faces, and 24 edges. It is made by cutting off the corners of a cube.

Notice that the triangular face of the cube octahedron IJK is formed by cutting off the corner of the cube G, and that the square face of the cube octahedron NPJI is formed when the 4 corners of the cube H, E, F, and G are cut off.

The cube octahedron has 8 triangular faces and 6 square faces. Figure 1 shows 3 of the square faces and 4 of the triangular faces.

There are 6 square faces on the cube octahedron, one for each face of the cube.

There are 8 triangular faces on the cube octahedron, one for each vertex of the cube.

This polyhedron has the fascinating property that the radius of the enclosing sphere, which touches all 12 vertices, is exactly equal to the length of all of the sides of the cube octahedron. Unfortunately, our 2 dimensional perspective cannot accurately capture the cube octahedron in true perspective, but if you build a model of one you'll see it's true.

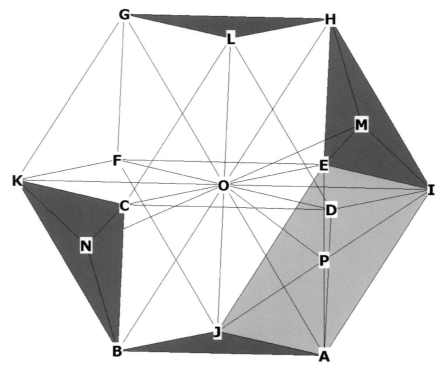

Fig. 10-1B

In Figure 10-1B you can see some of the rays coming out from O, for example OB and OA. All of these rays are equal in length to the sides, for example, GL and LC. This means that all of the rays branching out from the centroid, O, do so at 60 degree angles, forming 4 'great circle' hexagonal planes on the outside of the figure.

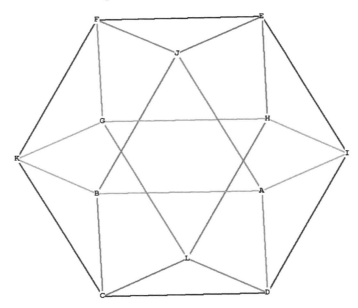

Fig. 10-2: showing the 4 'great circle' hexagons

The cube octahedron, if you observe Fig. 10-1B closely, can be seen to be composed of 8 tetrahedrons and 6 half-octahedrons. OGLH and OJAB, for example, are tetrahedrons. The half-octahedrons are formed from the square planes, for example OEIAJ and OCKGL.

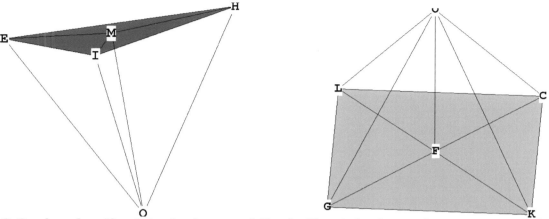

Fig. 10-3: showing the tetrahedron and the half-octahedron, the two types of pyramids of the cube octahedron.

The 12 vertices of the cube octahedron can also be considered to be composed of 3 orthogonal squares centered around O, the sides of which cross through the square faces of the cube octahedron as the diagonals of the squares:

Fig. 10-4 showing the 3 interlocking squares, the corners of which are the vertices of the cube octahedron. Note how the sides of the squares are the diagonals of the square faces of the cube octahedron.

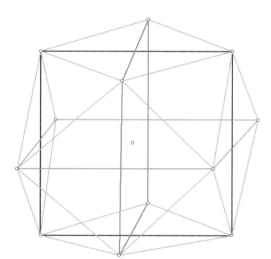

Fig. 10-4

Notice that in 10-5 at right, the intersection points of the squares form the 6 vertices of an octahedron:

The diagonals of the cube octahedral "squares" intersect to form the vertices of an octahedron (in purple)

What is the ratio of the radius of the enclosing sphere around the cube octahedron, and the side of the cube octahedron?

r = s.

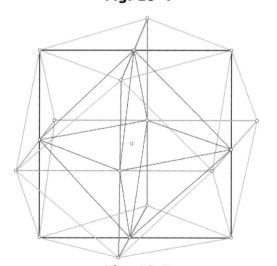

Fig. 10-5

All line segments from the centroid to any vertex are identical in length to the edge length's of the cube octahedron.

What is the volume of the cube octahedron (hereinafter referred to as c.o.)?

We use the pyramid method as usual.

V = 1/3 * area of base * pyramid height.

We have 8 triangular faces and 6 square faces.

From *The Equilateral Triangle* we know that the area of the triangular face is $\frac{\sqrt{3}}{4}s^2$.

The area of the square face is just s * s = s².

We need to find the height of the square pyramid and the height of the triangular pyramid.

In Figure 3, the triangular pyramid, or tetrahedron, is OEHI, with height OM. Triangle OMH is right, the line OM being perpendicular to the plane EHI at M. The height of the tetrahedral pyramid is known, from Tetrahedron, as $\frac{\sqrt{2}}{\sqrt{3}}s$.

In Figure 3, F is the center of the square face LCKG, and OF is the pyramid height. The triangle OFL is right, the line OF being perpendicular to the plane LCKG at F. OL is just the radius of the enclosing sphere, which, in the c.o., is the same as the c.o. side, s.

LF is one half the diagonal of the square LCKG, or $\frac{1}{\sqrt{2}}s$, because the diagonal of a square is always $\sqrt{2}$ times the side of the square.

Therefore we can write for the pyramid height OF, $\overline{OF}^2 = \overline{OL}^2 - \overline{LF}^2 = 1s^2 - \frac{1}{2}s^2 = \frac{1}{2}s^2$.

$OF = \frac{1}{\sqrt{2}}s$.   Note that this distance is identical to LF.

The Volume of 1 triangular pyramid then, is $\frac{1}{3} \times \frac{\sqrt{3}}{4}s^2 \times \frac{\sqrt{2}}{\sqrt{3}}s = \frac{\sqrt{2}}{12}s^3$.

The Volume of all 8 triangular pyramids is $8 \times \frac{\sqrt{2}}{12}s^3 = \frac{2\sqrt{2}}{3}s^3$.

The Volume of 1 square pyramid $= \frac{1}{3} \times s^2 \times \frac{1}{\sqrt{2}}s = \frac{1}{3\sqrt{2}}s^3$.

The Volume of all 6 square pyramids $= 6 * \frac{1}{3\sqrt{2}}s^3 = \frac{2}{\sqrt{2}}s^3$.

Therefore the Total Volume of the cube octahedron =

$\frac{2\sqrt{2}}{3}s^3 + \frac{2}{\sqrt{2}}s^3 = \frac{5\sqrt{2}}{3}s^3 = 2.357022604\,s^3 = 2.357022604\,r^3$.

This figure is larger than for the cube, but why? The volume of the cube in the unit sphere is only 1.539600718 r³ ! The reason lies in the fact that the cube octahedron fits more snugly within the unit sphere than does the cube.

The model on my desk shows the cube octahedron inside the cube, as in Fig.10-1A. Here, the side of the cube octahedron is $\frac{1}{\sqrt{2}} \times$ side of cube. This can be determined by an inspection of triangle GIJ in Figure 1.

GI = GJ = 1/2 the side of the cube. Angle IGJ is right. Therefore, GJ, the side of the cube octahedron = $\sqrt{\frac{1}{4} + \frac{1}{4}} = \frac{1}{\sqrt{2}} \times$ side of cube .

Calculating the volume of the cube octahedron in terms of the side of the cube as in Fig. 10-1A, we write

$$\text{Volume}_{cubeoctahedron} = \frac{5\sqrt{2}}{3} * \left(\frac{1}{\sqrt{2}}\right)^3 = \frac{5\sqrt{2}}{3} * \frac{1}{2\sqrt{2}} = \frac{5}{6} * soc^3,$$

where soc is the side of the cube.

Therefore the volume of the cubeoctahedron is 5/6ths the volume of the cube when the cubeoctahedron is sitting inside the cube.

There are 8 small tetrahedrons which represent the volume of the cube octahedron that have been "cut out" of the cube (See Fig. 10-1A and the tetrahedron GKIJ, for example). The volume of each of these tetrahedrons must then be one–eighth of 1/6, the difference between the volume of the cube and the volume of the cube octahedron. Therefore the volume of each small tetrahedron $= \frac{1}{8} \times \frac{1}{6} = \frac{1}{48} \times soc^3$, where soc is the side of the cube .

What is the surface area of the cube octahedron?

The surface area is 8 * area of triangular face + 6 * area of square face =

$$8 \times \frac{\sqrt{3}}{4} s^2 + 6 \times s^2 = (6 + 2\sqrt{3})s^2 = 9.464101615 \ s^2 \ .$$

What is the central angle of the cube octahedron? Each of the internal angles of the c.o. are formed from any of the 4 hexagonal planes which surround the centroid (see Fig. 10-2). Therefore the central angle of the cube octahedron = 60°.

There are 2 surface angles of the c.o., one being the 60° angle of the triangular faces, the other being the 90° angle of the square faces.

What is the dihedral angle of the cube octahedron? This angle is the intersection between a square face and a triangular face. If you sit the c.o. on one of its square faces you can see the following:

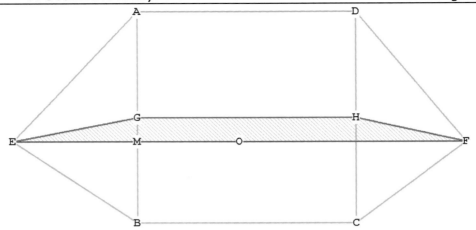

Fig. 10-6 - illustrating the dihedral angle of the cube octahedron -- top view

I have drawn the line EGHF to illustrate the dihedral angle between the two triangular faces and the square face ABCD. The point G is directly above the point M. O is the centroid of the c.o. M bisects EO.

The triangle GME is right by construction. The angle MGH is right by construction. The angle EGH is the dihedral angle. If we can find the angle EGM, all we have to do then is add 90° to it (∡MGH), and we have the dihedral angle.

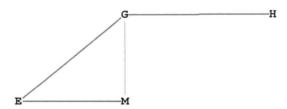

Fig. 10-7 -- showing that ∡ GME and ∡ MGH are right angles

OE = OF = side of c.o., or s. EM = 1/2 * OE.

EG is the height of a triangular face, which we know from The Equilateral Triangle is $\frac{\sqrt{3}}{2}$s .

So we can write:

$$\sin(\angle EGM) = EM / EG = \frac{\frac{1}{2}}{\frac{\sqrt{3}}{2}} = \frac{1}{\sqrt{3}} .$$

arcsin(1/√3) = 35.26438968° .

∡EGH = 90° + 35.26438968° = 125.26438968° .

Dihedral angle = 125.26438968° .

Of course, the dihedral angle between the square faces is 90°

The distance from the centroid to each vertex = s = r.

The distance from the centroid to any mid-edge is just the height of an equilateral triangle (remember that the centroid is surrounded by 4 hexagons (Fig. 10-2). This distance is $\frac{\sqrt{3}}{2}$ s.

The distance from the centroid to a triangular mid-face is just the height of the tetrahedron OM

(see Fig. 10-3), which we know from above, as $\frac{\sqrt{2}}{\sqrt{3}}$ s.

The distance from the centroid to a square mid-face is just the height of the half–octahedron OF

(see Fig. 10-3), which we know from above, is $\frac{1}{\sqrt{2}}$ s.

The relationship between these distances is 1, 0.866025404, 0.816496581, 0.707106781

Cube Octahedron Reference Tables

Volume (edge)	Volume in Unit Sphere	Surface Area (edge)	Surface Area in Unit Sphere
2.357022604 s^3	2.357022604 r^3	9.464101615 s^2	9.464101615 r^2

Central Angle:	Dihedral Angle:	Surface Angles:	
60°	125.26438968°(between square and triangular faces) 90° (between square faces)	60°	
		90°	

Centroid To: Vertex	Centroid To: Mid–Edge	Centroid To: Mid–triangle Face	Centroid To: Mid–square Face
1.0 r	0.866025404r	0.816496581r	0.707106781r
1.0 s	0.866025404s	0.816496581s	0.707106781s

Side / radius
1.0

Chapter 11 – The Star Tetrahedron

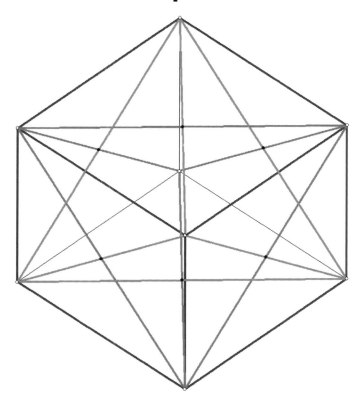

Figure 11-1A. The Star Tetrahedron in the Cube

Fig. 11-1A shows the cube in blue, and two interlocking tetrahedrons, one in purple, the other in green. To visualize this, imagine that the green tetrahedron comes out of the page at the top front corner of the cube and the purple tetrahedron goes back toward the bottom corner of the cube, into the page.

The two interlocking tetrahedrons each intersect the other by bisecting each others' sides. This becomes much clearer when you build a 3D model.

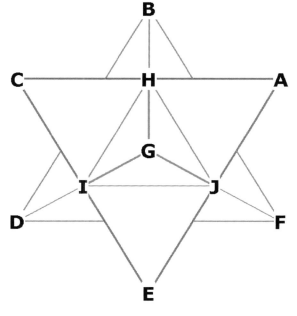

Fig. 11-1B

Fig. 11-1B shows that the interlocking tetrahedra bisect the sides of each other. BG bisect CA at H, FG bisects AE at J, DG bisects CE at I. There are 12 interlocking edges, 6 for each tetrahedron.

What is the volume occupied by a star tetrahedron inside a cube? The intersection of the two interlocking tetrahedrons forms an octahedron plus 8 smaller tetrahedron's that stick out from the octahedron, as shown on Fig.11-1C.

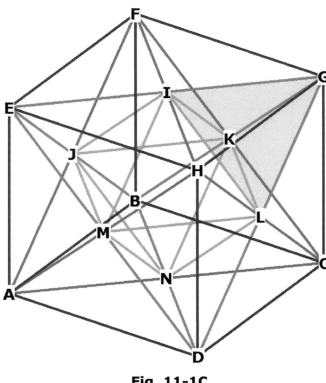

Fig. 11-1C

Showing the 8 smaller tetrahedron that stick out from the octahedron, one from each of the octahedral faces. I have shaded one of these tetrahedra (GILK).

The volume of the star tetrahedron is the volume of the octahedron + volume of each small tetrahedron, so we write:

$$\text{Volume}_{\text{star tetrahedron}} = \text{Volume}_{\text{octahedron}} + \text{Volume}_{\text{8 small tetrahedrons}}$$

Let cs = side of the cube = 1. Then $V_{\text{cube}} = 1^3 = 1$.

The length of the edges of the interlocking tetrahedrons are each

$\sqrt{2}$ * side of the cube, because each of them is a diagonal of a square face of the cube.

The edges of each of the small tetrahedrons are exactly 1/2 side of the large tetrahedron, because each of the large tetrahedrons bisects the other in order to form the interior octahedron. That is the magic of the Platonic Solids!

So the sides of the small tetrahedron = $\dfrac{\sqrt{2}}{2}$ cs.

The sides of the interior octahedron are also $\dfrac{\sqrt{2}}{2}$ cs , because the vertices of the octahedron bisect the edges of the interlocking tetrahedrons. This becomes clear when you build a three dimensional model.

We know from Octahedron that the volume of an Octahedron is $\dfrac{\sqrt{2}}{3}$ * side of octahedron (os).

We need to convert this to the side of the cube:

$$V_{\text{octahedron}} = \frac{\sqrt{2}}{3} * \left(\frac{\sqrt{2}}{2}\,\text{cs}\right)^3 = \frac{\sqrt{2}}{3} * \frac{2\sqrt{2}}{8}\,\text{cs}^3 = \frac{1}{6}\,\text{cs}^3 .$$

We know from Tetrahedron that the volume of a Tetrahedron

is $\dfrac{1}{6\sqrt{2}} \times ts^3$, where ts = side of tetrahedron . It turns out that the edge lengths of each of our 8

tetrahedrons is equal to the lengths of the sides of the octahedron. So we may write

$$V_{\text{small tetrahedron}} = \frac{1}{6\sqrt{2}} \times \left(\frac{\sqrt{2}}{2} cs\right)^3 = \frac{1}{6\sqrt{2}} \times \frac{\sqrt{2}}{4} cs^3 = \frac{1}{24} cs^3.$$

$$V_{\text{total, 8 small tetrahedrons}} = 8 \times \frac{1}{24} cs^3 = \frac{1}{3} cs^3.$$

$$V_{\text{total, star tetrahedron}} = \frac{1}{6} cs^3 + \frac{1}{3} cs^3 = \frac{1}{2} cs^3.$$

Even though the volume of each large tetrahedron in the star tetrahedron is exactly 1 / 3 of the cube, the star tetrahedron only occupies one half the volume of the cube.

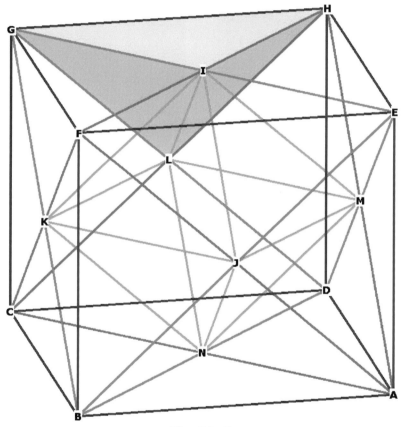

Fig.11-2.

There are 8 congruent solids (for example, GLHI) representing the left over space in the cube not included in the star tetrahedron volume.

Since the left over space in the cube $= \dfrac{1}{2} cs^3$, $\quad$ Volume$_{1 \text{ left over solid}} = \dfrac{1}{8} * \dfrac{1}{2} cs^3 = \dfrac{1}{16} cs^3.$

These solids have 4 vertices, 4 faces, and 6 edges, thus fulfilling the Euler requirement that Faces + Vertices = Edges + 2.

The Star Tetrahedron in the Sphere

The sphere enclosing the star tetrahedron is the same sphere that encloses the cube.

The diameter of the enclosing sphere is the distance marked in orange in Figure 4, with the centroid of the star tetrahedron marked with a black dot:

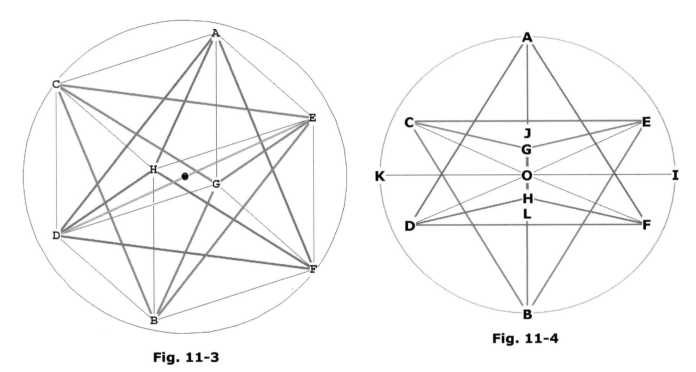

Fig. 11-3

Fig. 11-4

Fig.10-4 shows a different view of the star tetrahedron All 8 points of the star tetrahedron touch the surface of the enclosing sphere. Note that the diameter of the enclosing sphere is the same diagonal that goes through the center of the cube. In Figure 3 this is marked as ED. From Cube we know that the length of this diagonal, and the diameter of the enclosing sphere, is $\sqrt{3}$. (This calculation is a simple one, ED being the hypotenuse of the right triangle EDF. Since DF is the diagonal of the cube with sides = 1, DF = $\sqrt{2}$. FE = 1, because it's the side of the cube, so by the Pythagorean Theorem, ED = $\sqrt{3}$).

There are 4 axes of rotation for the star tetrahedron; AB, CF, DE, and GH.

Notice the planes CGE in green and HDF in purple; both planes go into and out of the screen or page.

What is the distance between these planes? This distance will be the line JOL. J lies at the center of CGE, and L lies at the center of HDF. O is the centroid. By construction, both planes are parallel to each other.

What is the angle between the xy plane at the origin and the planes CGE and HDF? This will be $\measuredangle$ IOE, for example.

It turns out that if you have a model of the star tetrahedron you can easily see that the distance between the two planes, JOL, is the height of one of the smaller tetrahedrons from the previous section.

We know that the height of a tetrahedron is $\dfrac{\sqrt{2}}{\sqrt{3}}$ * side of tetrahedron

$$= \frac{\sqrt{2}}{\sqrt{3}} \times \frac{\sqrt{2}}{2} \times \text{side of cube} = \frac{1}{\sqrt{3}} \times \text{side of cube} \,.$$

JOL $= \dfrac{1}{\sqrt{3}}$ cs. This distance is exactly 1/3 of the diameter of the enclosing sphere.

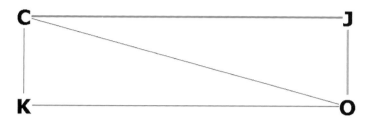

Fig. 11-5. Showing JO, the distance between the plane CGE and the centroid O.

JO = one half of JOL $= \dfrac{1}{2\sqrt{3}}$ cs.

CJ represents the plane CGE and OK represents the plane DFH. J is the center of the plane CGE, and L is the center of the plane DFH.

CO is the radius of the enclosing sphere, and equal to $\dfrac{\sqrt{3}}{2}$ cs.

Now we have OJ and CO and can get the angle COJ:

$$\cos(\angle COJ) = OJ / OC = \frac{\left(\dfrac{1}{2\sqrt{3}}\right)}{\left(\dfrac{\sqrt{3}}{2}\right)} = \frac{1}{3}\text{, so}$$

$\angle COJ = 70.52877937°$ and
$\angle COK = 19.47122063°$

Chapter 12 – The Rhombic Dodecahedron

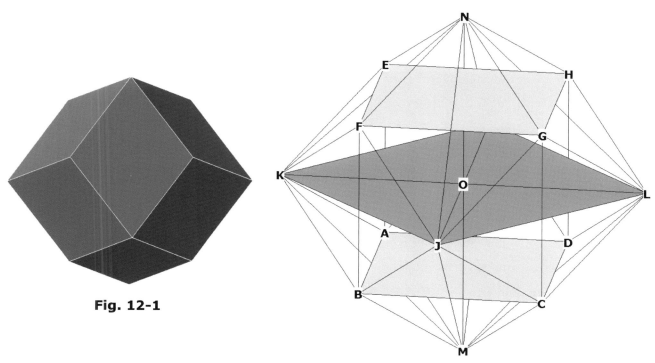

Fig. 12-1

The rhombic dodecahedron is a very interesting polyhedron. It figures prominently in Buckminster Fuller's *Synergetics*. It has 12 faces, 14 vertices, 24 sides or edges.

Section 1 – Introduction

Fig. 12-1A shows the edges of rhombic dodecahedron (yellow), octahedron (green) and cube (blue).

Notice in Figures 12-1 and 12-1A that the rhombic dodecahedron is composed of diamond faces (for example, NFJG).

The faces are called rhombuses, because they are equilateral parallelograms. In other words, they are square-sided figures with opposite edges parallel to one another. The rhombic dodecahedron has 8 vertices in the middle that form a cube, (ABCD-EFGH), and the other 6 on the outside which form an octahedron (N-ILJK-M).

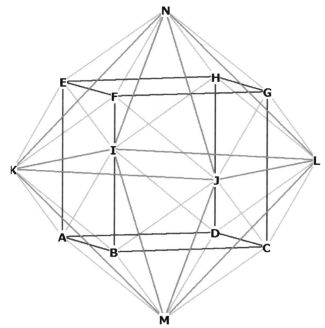

Fig. 12-1A: Same as Fig. 12-1
slightly rotated

It is possible to draw a sphere around the cube, and another, larger sphere, around the octahedron. Unlike the 5 regular solids, therefore, not all of the vertices of the rhombic dodecahedron
will touch one sphere. We will, in the course of this analysis, find the diameter and radius of
each of these spheres.

In Fig. 12-1, I have marked the top and bottom planes of the cube in light gray, and the square
plane which serves as the base for the 2 face-bonded pyramids of the octahedron, in dark gray.
You may perceive the edges of the cube in this drawing, which connect the top and bottom
planes; for example, FB,GC, HD; and also some of the edges of the octahedron, the most visible of which are NJ, NK, NL, MK, MJ, ML. Notice that the edges of the octahedron bisect the
diamond faces of the rhombic dodecahedron upon their long axis (for example, NK bisects the
long axis of the face NEKF at the upper left). Note also that the short axis segments (that is, EF)
are the edges of a cube.

It is important to understand that the outer vertices of the rhombic dodecahedron form an octahedron, as this will be an important part of the analysis.

The rhombic dodecahedron (hereinafter, referred to as r.d.) is a semi-regular polyhedron, in
that all of its edges are the same length, yet the angles of its faces differ. Because each face is a
parallelogram, there are 2 distinct angles for each face, one which is bisected by the long axis,
with an angle less than 90 degrees, the other bisected by the short axis, with an angle greater
than 90 degrees. Of course these axes do not actually appear on the face of the polyhedron, I
use them here for illustration.

In Fig. 12-2, the vertices of the octahedron are the long axis vertices of each face, in this case N
and J. The cube vertices are the short axis points, in this case, F and G. This can be more clearly
seen by referring to Figure 1.

Fig. 12-2 shows the general appear-
ance of the r.d. The front 4 faces of
the octahedron can be seen clearly
here in green (NKJ, NJL, KJM and
JLM). Figure 2 also shows how the
short axis vertices of the r.d. (as F, G,
B, C) come off the face of the octahedron. The long axis vertices (as in N,
J) are vertices of the octahedron. O is
the centroid and is beneath J.

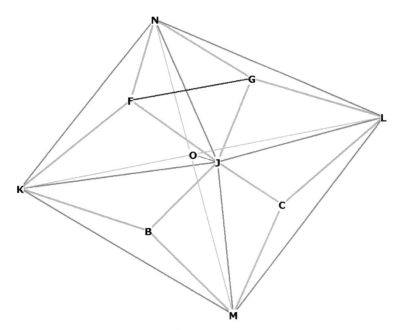

Fig. 12-2

The sides of the rhombic dodecahedron are JFNG, one of the sides of the octahedron is JN. $\angle$ JON is the central angle of the long axis of the r.d. face, and $\angle$ FOG is the central angle of the short axis. O is the centroid of the r.d. and of the octahedron and cube within the r.d. When you build a 3D model of the r.d., it appears that OF = OG = NF = FJ = JG = NG by construction; that is, the distance from the centroid O to any of the 6 short axis vertices of the r.d. faces (the vertices that make the cube, see Fig. 12-1A) are equal in length to the edges of the rhombic dodecahedron. We will prove this later on.

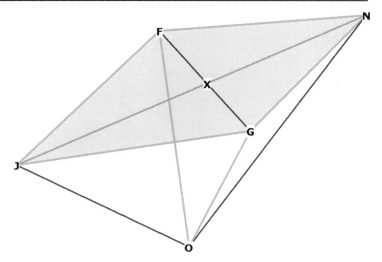

Figure 12-3

Fig. 12-4 shows the point F, one of the vertices of the rhombic dodecahedron, off one of the faces of the octahedron, NJK. The centroid of the octahedron/r.d. is at O. Q is the center of the octagonal face NJK.

If you build a model of the rhombic dodecahedron with the Zometool, you will see at once that the sides of the r.d. come off any of the faces of the octahedron and meet above the center of the octahedron face (for example, at F). F is also the centroid of a tetrahedron with side length = side of the octahedron that can be formed from the face NJK of the octahedron (see Fig. 12-4A below). Tetrahedrons may be formed from any of the faces of the octahedron, and connected in the fashion of Fuller's Isotopic Vector Matrix.

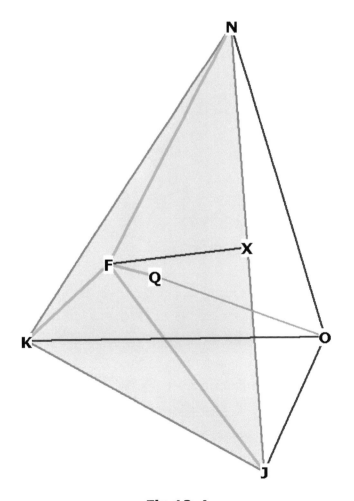

Fig.12-4

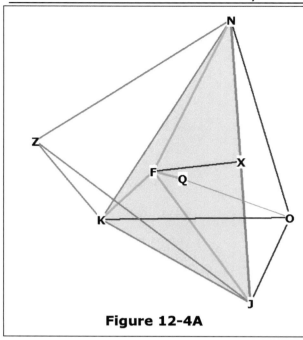

Figure 12-4A

QF, in Fig. 12-4A, is the distance from the plane of the octahedron/tetrahedron face, to the centroid of any such tetrahedron. XF is the distance from the center of any of the diamond faces of the r.d. to a short axis vertex, in this case, F.

OQ is the distance from the centroid to the middle of the face of the octahedron. The centroid of the octahedron is also the centroid of the r.d. (O), which is built around the faces of the octahedron, as shown in Fig. 12-1.

Let's find the volume of the rhombic dodecahedron. As usual, we will use the pyramid method.

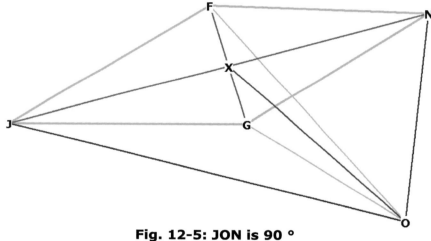

Fig. 12-5: JON is 90 °

The height of any pyramid is 1/3 x area of base x height of pyramid.

We need to get the area of each diamond face.

In order to do this we need to take a rather extensive detour in which we will derive lots of interesting information about the rhombic dodecahedron.

Section 2 – Rhombic Dodecahedron Internals

First off, we need to know the length of the side of the rhombic dodecahedron. In reference to what? We have an octahedron and a cube inside the r.d. We already have the lengths of each of these sides, in relation to a sphere that encloses the octahedron and cube. The outer sphere of the r.d. is precisely that sphere that touches all 6 vertices of the octahedron, so we choose the edge of the octahedron as our reference point.

Recall from *Octahedron* that this relationship is: $r = \dfrac{1}{\sqrt{2}} \times$ side of octahedron .

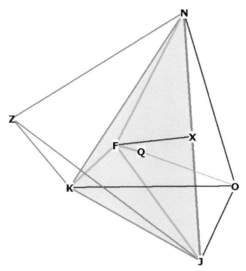

Fig. 12-4A, repeated

The distance FX in Fig. 12-4A is the distance from the centroid of the tetrahedron to the mid–

edge of the tetrahedron. We know from Tetrahedron that this distance $= \dfrac{1}{2\sqrt{2}} *$ side of the tet-

rahedron. Therefore, FX $= \dfrac{1}{2\sqrt{2}} *$ side of the octahedron.

Now we can find the side of the r.d. in terms of the octahedron side.

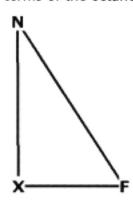

Fig.12-6 (Refer also to Figs. 12-4A and 12-5).

NX = 1/2 the side of the octahedron, or long-axis of the r.d. face.

NF is the r.d. side.

We will refer to "os" as the side of the octahedron. "os" is equal in length to the side of a tetra-

hedron whose face is congruent to the octahedron face (See Fig. 12-4A).

"rds" will be the r.d. side.

$$\overline{NF}^2 = \overline{NX}^2 + \overline{FX}^2 = \frac{1}{4}os^2 + \frac{1}{8}os^2 = \frac{3}{8}os^2.$$

$$NF = rds = \frac{\sqrt{3}}{2\sqrt{2}}os = \frac{3}{2\sqrt{6}}os.$$

$$os = \frac{2\sqrt{2}}{\sqrt{3}}rds = \frac{2\sqrt{6}}{3}rds.$$

While we're at it, lets get FG, the short-axis distance on the r.d. face. FG is visible in Figures

1,2,3 and 5. We have already established $FX = \frac{1}{2\sqrt{2}}os$. Substituting,

$$FX = \frac{1}{2\sqrt{2}} * \frac{2\sqrt{2}}{\sqrt{3}}rds = \frac{1}{\sqrt{3}}rds.$$

Since FG = 2 * FX, $FG = \frac{2}{\sqrt{3}}rds.$

It is important to establish the distance OF = OG, the distance from the r.d. centroid to the short axis points on the r.d. face. We have shown these in yellow (Fig.12-3 and Fig. 12-5), indicating their length is equal to the length of the r.d. side. Is this true?

Triangle OXF is right, by construction. In order to get OF = OG, we need OX, which also happens to be the height of the r.d. pyramid.

The height of the pyramid can be determined by inspection. From Figures 2 and 4 we see that OX goes from the centroid of the r.d. to the midpoint of the octahedron side. But this distance is exactly 1/2 the octahedron side, as we see clearly in Figure 7.

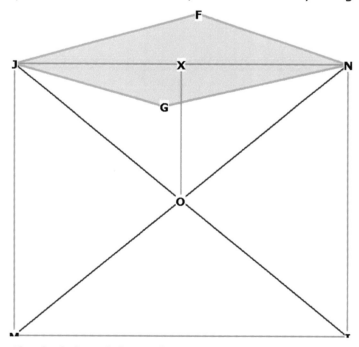

Fig. 12- 7, showing the height of the r.d. pyramid (OX) is 1/2 os, the side of the octahedron (NI, JM).

So h = OX = $\frac{1}{2}$ os, and os = $\frac{2\sqrt{2}}{\sqrt{3}}$ rds, therefore h = $\frac{1}{2} * \frac{2\sqrt{2}}{\sqrt{3}} = \frac{\sqrt{2}}{\sqrt{3}}$ rds.

Now we have the height of the r.d. pyramid in terms of the r.d. side.

$$XN = \frac{1}{2} os = \frac{1}{2} * \frac{2\sqrt{2}}{\sqrt{3}} rds = \frac{\sqrt{2}}{\sqrt{3}} rds .$$

The triangle OXN is isosceles.

Now that we have calculated OX, it remains to prove that OF = OG. When you build a model of the rhombic dodecahedron you can see this immediately, but lets show it mathematically as well. We will find the result first in terms of the side of the octahedron, then translate this with respect to the side of the r.d.

$$\overline{OF}^2 = \overline{OG}^2 = \overline{OX}^2 + \overline{XF}^2 = \frac{1}{4} os^2 + \frac{1}{8} os^2 = \frac{3}{8} os^2$$

$OF = OG = \frac{\sqrt{3}}{2\sqrt{2}} os$, which is precisely what we established above.

$OF = OG = \frac{\sqrt{3}}{2\sqrt{2}} * \frac{2\sqrt{2}}{\sqrt{3}} rds = rds$

Therefore OF = OG is equal in length to the side of the r.d.

What is the relationship between OQ and FQ? From Tetrahedron and Octahedron we know that the distance from the centroid of the tetrahedron to the mid-face = $\frac{1}{2\sqrt{6}} os = FQ$.

The distance from the centroid of the octahedron to the mid-face =

$\frac{1}{\sqrt{6}} os = OQ$.

Therefore OQ = 2 x FQ, and we conclude that the face of the octahedron is twice as far away from its centroid, as is the face of the tetrahedron from its centroid.

We also gain the important information that triangle FOG is isosceles.

Now we can get to some important data, that is, the relationship between the side of the r.d. and the radius of the sphere that contains the rhombic dodecahedron. That sphere is, you re-call, the sphere that surrounds and touches the vertices of the octahedron (N-ILJK-M in Figure 1 and 1A). We will refer to this radius as r_{outer} .

The radius of this sphere is ON = OJ in Fig. 12-8 below. What is ON = OJ?

∡ JON is right by construction.

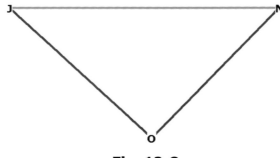

Fig. 12-8

We know that JN = side of the octahedron = os. OJ = ON = r_{outer}, so

$$2 * r_{outer}^2 = os^2, \quad r_{outer}^2 = \frac{os^2}{2}.$$

$$r_{outer} = \frac{1}{\sqrt{2}} os.$$

We found that $rds = \frac{\sqrt{3}}{2\sqrt{2}} os$, so we substitute for os and get,

$$r_{outer} = \frac{1}{\sqrt{2}} * \frac{2\sqrt{2}}{\sqrt{3}} rds = \frac{2}{\sqrt{3}} rds.$$

Therefore $r_{outer} = \frac{2}{\sqrt{3}} rds$, and $rds = \frac{\sqrt{3}}{2} r_{outer}$.

Now we have expressed the side of the r.d. in terms of the radius of the sphere that contains the rhombic dodecahedron, realizing that this sphere touches only the 6 outer vertices of the r.d.

Notice that FG, the short-axis distance across the face of the r.d., (calculated above), also

$$= \frac{2}{\sqrt{3}} rds.$$

We also know from Figure 3 that the radius of the smaller sphere touching the 6 cube vertices ABCD-EFGH is equal to the side length of the r.d. We may write $rds = r_{inner}$ for this smaller sphere.

(Henceforth, we will write "r" for the radius, understanding that $r = r_{outer}$).

Section 3 – The Resumption of the Volume Calculation

We were, at the end of Section 1, about to find the volume of the rhombic dodecahedron. To do this we need to recall the height of the pyramid OX (see Fig. 12-7), and find the area of the r.d. face.

By looking at Figure 3 we perceive that the area of the r.d. face can be divided into 2 triangles, each one with base NJ. Both triangles are congruent by side-side-side, and the height, ht, of each is XF = XG.

From above we know that this distance is $\frac{1}{2\sqrt{2}}$ os.

NJ is the side of the octahedron, so NJ = os.

Now we have

$$\text{area}_{1 \text{ face triangle}} = \frac{1}{2} * \text{base} * \text{height} = \frac{1}{2} * \text{os} * \frac{1}{2\sqrt{2}} \text{os} = \frac{1}{4\sqrt{2}} \text{os}^2.$$

$$\text{area}_{1 \text{ face}} = 2 * \text{area}_{1 \text{ face triangle}} = \frac{1}{2\sqrt{2}} \text{os}^2.$$

We want to get all of our data on the r.d. in terms of the r.d. itself, for consistency. Therefore we translate the area from the side of the octahedron, to the side of the r.d.

$$\text{area}_{1 \text{ face}} = \frac{1}{2\sqrt{2}} \text{os}^2 = \frac{1}{2\sqrt{2}} * (\frac{2\sqrt{2}}{\sqrt{3}})^2 \text{rds}^2 = \frac{1}{2\sqrt{2}} * \frac{8}{3} \text{rds}^2$$

$$= \frac{4}{3\sqrt{2}} \text{rds}^2 = 0.942809042 \text{ rds}^2.$$

Now, finally, we have enough data to calculate the volume of 1 pyramid:

$$\text{Volume}_{1 \text{ pyramid}} = \frac{1}{3} * \text{area of base} * h = \frac{1}{3} * \frac{4}{3\sqrt{2}} \text{rds}^2 * \frac{\sqrt{2}}{\sqrt{3}} \text{rds}$$

$$= \frac{4}{9\sqrt{3}} \text{rds}^3$$

$$\text{Volume}_{\text{r.d.}} = 12 * \text{Volume}_{1 \text{ pyramid}} = 12 * \frac{4}{9\sqrt{3}} \text{rds}^3$$

$$\text{Volume}_{\text{r.d.}} = \frac{16}{3\sqrt{3}} \text{rds}^3 = 3.079201436 \text{ rds}^3.$$

What is the surface area of the rhombic dodecahedron?

It is just 12 faces * area of 1 face = $12 * \frac{4}{3\sqrt{2}} \text{rds}^2 = \frac{16}{\sqrt{2}} \text{rds}^2 = 8\sqrt{2} \text{rds}^2.$

$$\text{Surface area}_{\text{rhombic dodecahedron}} = 8\sqrt{2} \text{ rds}^2 = 11.3137085 \text{ rds}^2.$$

There are 3 central angles of the rhombic dodecahedron.

In Figure 3, they would be, for example, $\angle$ FOG, $\angle$ FON, $\angle$ JON.

Let's start with $\angle$ FOG.

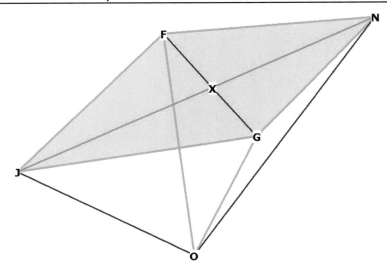

Fig. 12-3, repeated

∡ JON is right. ∡ OXF, ∡ OXG are right.

To find ∡ FOG, find ∡ FOX. Triangle FOX is right by construction, therefore,

$$\sin(\angle FOX) = FX / FO = \frac{\frac{1}{\sqrt{3}}}{1} = \frac{1}{\sqrt{3}},$$

$$\angle FOX = \arcsin(\frac{1}{\sqrt{3}}) = 35.26438968°,$$

$$\angle FOG = 2 * \angle FOX = 70.52877936°$$

This is precisely the central angle of the cube! Understandably so, for we already know that the two vertices F and G are two of the vertices of a cube (see Fig. 12-1A).

It is also the dihedral angle of the tetrahedron.

Because the triangle FOG is isosceles,

∡ OFG = ∡ OGF = ∡ FON = (180° - 70.52877936°) / 2, using the property that the sum of angles of a triangle = 180° .

Therefore ∡ FON = 54.7356103° (and so does ∡ FNO).

From Fig. 12-7 we can see immediately that ∡ JON is right, it being the central angle of the square JMIN. ∡ JON = 90°

What are the surface angles of the diamond faces of the rhombic dodecahedron?

We know that the short-axis distance across the r.d. face,

FG in Figure 3, is $\frac{2}{\sqrt{3}}$ rds .

We know FN = NG = the side of the r.d. = rds.

Triangle FXN is right by construction.

∡ FNX is one half the face angle FNG. FX is one-half FG.

So we can write:

$$\sin(\angle FNX) = FX / FN = \frac{1}{\sqrt{3}}.$$

$\angle FNX = 35.26438968°$, so

$\angle FNG = 70.52877936°$.

$\angle FNG$ and its counterpart $\angle FJG$ are the smaller of the two face angles of the r.d. Now lets get

$\angle JFN$.

Once again we use triangle FXN and work with $\angle XFN$, which is one-half the desired angle, JFN.

$\angle XFN$ is just 90 - $\angle FNX$, using the property that the sum of all angles in a triangle is 180°.

$\angle XFN = 54.73561032°$, so

$\angle JFN = 109.4712206°$.

This angle, 109.4712206°, is the central angle of the tetrahedron and the dihedral angle of the octahedron. It also is the angle you see when you stand the rhombic dodecahedron on one of its 8 octahedral vertices and look down from above at two intersecting r.d. sides. This polyhedron an all-space filler, meaning that it can be joined with itself, like the cube, to fill any volume without any space left over.

If you examine the r.d. from the inside out with the *zometool*, you will see it is composed internally of the same rhombi as the faces.

To identify these internal rhombuses, lay the rhombic dodecahedron flat on one of its faces. Find 3 r.d. points and the centroid to see the rhombus:

There are 12 internal rhombi, and 12 external rhombi, each of them identical. Note that there is another angle on the exterior of the rhombic dodecahedron, and that is the angle that one face plane makes with another face plane as it goes over top of the octahedron within the r.d. This angle is 90°. This can be seen by looking at the long axis diagonals of the octahedron through the rhombi, and following the faces that form around the side of the octahedron. In Fig. 12-1, this can be seen with the faces NFJG and JBMC.

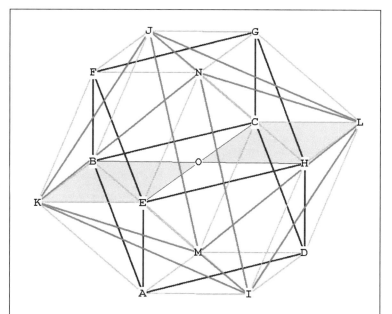

Fig. 12-9 – two of the 12 internal rhombi of the rhombic dodecahedron. O is the centroid.

Notice the lines NJ and JM, which are part of the octahedral square, and which bisect the faces along their long axis. $\angle$ NJM is a right angle, as is $\angle$ NLM and $\angle$ NKM. Observe also $\angle$ KNL.

Again, this property helps the rhombic dodecahedron to be an all-space filler.

Now lets complete the analysis of the rhombic dodecahedron by finding or collecting the distances from the centroid to a short-axis vertex, a long-axis vertex, mid-edge and mid-face. We already have all of this information, except for the distance from the centroid to a mid-edge:

distance from centroid to long-axis r.d. vertex (vertex of the octahedron) $= \dfrac{2}{\sqrt{3}}$ rds .

distance from centroid to short-axis r.d. vertex (vertex of the cube) = rds.

distance from centroid to r.d. mid-face (mid-point of octahedron side) $= \dfrac{\sqrt{2}}{\sqrt{3}}$ rds.

To find centroid to r.d. mid-edge requires a little work:

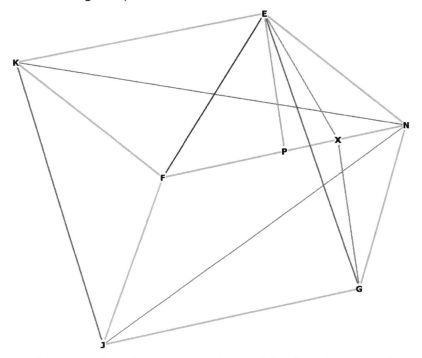

Fig. 12-9 -- showing centroid to mid-edge distance EP.

We have already remarked upon one of the properties of the r.d., that it is composed of 12 internal rhombi identical to the external faces. Those rhombi are divided in half (here, at EF) by the same distance as from the centroid to any of the vertices of the octahedron contained within the r.d. Therefore the r.d. is composed of 24 exterior congruent triangles (here, $\triangle$ EFN and $\triangle$ EKF), and 24 interior congruent triangles congruent to the exterior ones!

So triangle EFN may substitute for any triangle in the interior of the r.d., and more to the point, the vertex E or F may substitute for the centroid of the r.d., at O. In order to see this clearly, you have to build a 3D model of the rhombic dodecahedron, and I encourage you to do this.

From Rhombic Dodecahedron Dihedral Angle, we know that EX $= \dfrac{2\sqrt{2}}{3}$ rds.

We know that $\angle$ EXN is right by construction, and that EN = rds. We also know that P is at mid-edge on FN, the side of the r.d., therefore NP = 1/2 rds.

We can find XN, and so XP, by writing XP = NP - XN.

We need to find XN.

$$\overline{XN}^2 = \overline{EN}^2 - \overline{EX}^2 = 1rds^2 - \frac{8}{9}rds^2 = \frac{1}{9}rds^2.$$

$$XN = \frac{1}{3}rds.$$

$$XP = \frac{1}{2}rds - \frac{1}{3}rds = \frac{1}{6}rds.$$

Now we can find EP, the distance from centroid to mid-edge.

$$\overline{EP}^2 = \overline{EX}^2 + \overline{XP}^2 = \frac{8}{9}rds^2 + \frac{1}{36}rds^2 = \frac{33}{36}rds^2 = \frac{11}{12}rds^2.$$

$$EP = \frac{\sqrt{11}}{2\sqrt{3}}rds = 0.957427108 \; rds.$$

The Dihedral Angle of the Rhombic Dodecahedron

When calculating dihedral angles, it is vital to ensure that the angle chosen is an accurate representation of the intersection of the 2 planes. In the rhombic dodecahedron (r.d.), this is a little tricky.

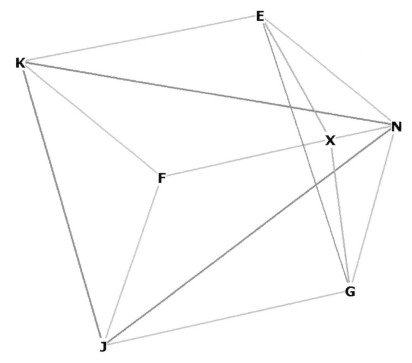

Fig. 12-10 - The r.d. dihedral angle

The dihedral angle of the r.d. is the angle EXG. This angle is a geometrically accurate description of the intersection of any 2 planes of the r.d., in this case, the 2 planes KENF and JGNF. The lines EX and GX are constructed such that ∡ NXG and ∡ NXE are right. ∡ EXG is a "roof" over the 2 planes which exactly describes the dihedral angle.

We will be working with triangle EXG .

It's easier to see this if you construct a model of the rhombic dodecahedron, and lay it on one of

its edges.

I have shown EG in green, indicating it is equal to the side of the octahedron (referred to as os).

We have to show this.

In Fig. 12-11, O is the centroid.
NKJ is one of the faces of the
octahedron. ON, OK and OJ are
radii of the outer sphere that
touch the 6 octahedral vertices.
OE and OG touch the short axis
vertices of the r.d.

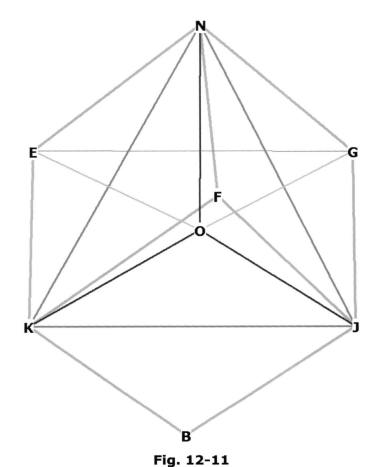

Fig. 12-11

Fig. 12-12 is another view showing how EG is found. NEOG is
one of 12 internal rhombi of the
rhombic dodecahedron.
OE and OG are the sides of the
rhombic dodecahedron, or rds.
We have previously shown that
the triangles OEN and OGN are
isosceles and congruent (the
vertices may have different
names, but the triangles are the
same!) NZE is right.

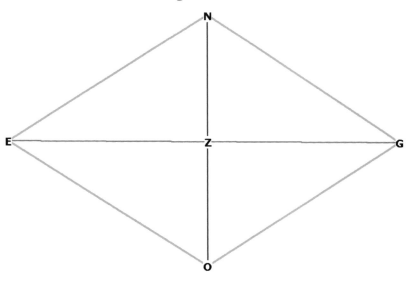

Fig. 12-12

To obtain EG, simply re-do the calculation we did previously,

$$\vec{EZ}^2 = \vec{EN}^2 - \vec{NZ}^2,$$

$$\vec{EZ}^2 = rds^2 - \frac{1}{3}rds^2 = \frac{2}{3}rds^2$$

$$EZ = \frac{\sqrt{2}}{\sqrt{3}} \ rds.$$

$$EG = 2 * EZ, \quad EG = \frac{2\sqrt{2}}{\sqrt{3}}rds$$

which has been obtained previously as os. Therefore EG is identical in length to the side of the octahedron.

This internal rhombus, NEOG, is identical to the rhombus of the r.d. face. The r.d. is composed, internally and externally, of identical rhombi, which gives it the quality of an all-space filler.

We have established EG as the side of the octahedron. Now it remains to determine GX and XE. Look at Fig. 12-10 again. The triangle ENX is right by construction, and we know the angle ENX, it being the long-axis angle of the r.d. face, as $2 * \arcsin(\frac{1}{\sqrt{3}})$, or 70.52877936 degrees.

We also know EN = side of the r.d, or rds.

So we write $\sin(\angle ENX) = EX / EN = EX / rds$.

$$EX = rds * \sin(70.52877936) = \frac{2\sqrt{2}}{3} \ rds = 0.942809042 \ rds.$$

(Note: EX = $\frac{1}{\sqrt{3}}$ os).

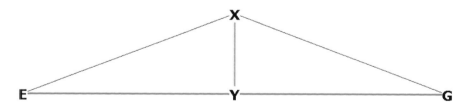

Fig. 12-13 -- showing the dihedral angle in simplified fashion

Now, EX and EY are known. We need to get $\angle EXY$.

We know EG is the side of the octahedron, or os, and it is $\frac{2\sqrt{2}}{\sqrt{3}}$ rds and so EY is $\frac{1}{2}$ that.

$$\sin(\angle EXY) = EY / EX = \frac{\frac{\sqrt{2}}{\sqrt{3}}rds}{\frac{2\sqrt{2}}{3}rds} = \frac{3}{2\sqrt{3}} = \frac{\sqrt{3}}{2}.$$

$$\angle EXY = \arcsin(\frac{\sqrt{3}}{2}) = 60°$$

Therefore $\angle$ EXG = 2 * $\angle$ EXY = 120°

Dihedral angle of the rhombic dodecahedron = 120°

Rhombic Dodecahedron Reference Charts

Volume (edge)	Volume in Unit Sphere	Surface Area (edge)	Surface Area in Unit Sphere
3.079201436 s³	2.0 r³	11.3137085 s²	8.485281375 r²
	Distance to outer sphere touching 6 octahedral vertices		Distance to outer sphere touching 6 octahedral vertices

Central Angles:	Dihedral Angle:	Surface Angles:
Long axis:	120°	Long axis:
90°		109.4712206°
Short axis:		Short axis:
70.52877936°		70.52877936°
Adjacent Vertex:		
54.7356103°		

Centroid To: Vertex	Centroid To: Mid–Edge	Centroid To: Mid–Face
1.0 r	0.829156198 r	0.707106781 r
1.154700538 s	0.957427108 s	0.816496581 s

Side / radius To 8 cube vertices:
1.0
To 6 octahedral vertices:
0.866025404

Chapter 13 – The Icosa–Dodecahedron

Fig. 13-1 The Icosadodecahedron

This polyhedron is the dual of the rhombic triacontahedron.

It has 30 vertices, 32 faces, and 60 edges.

20 of the faces are equilateral triangles. 12 of the faces are pentagons.

It is probably called the icosa-dodecahedron because in the middle of every pentagonal face is the vertex of an icosahedron, and in the middle of every triangular face is the vertex of a do-decahedron.

This polyhedron is composed of 6 'great circle' decagons which traverse the outside of the poly-hedron, sharing the 30 vertices and accounting for the 60 edges (see Fig. 13-2).

Actually, the rhombic triacontahedron should be called the icosa dodecahedron, because one icosahedron and one dodecahedron precisely describe its 32 vertices.

Fig.13-2 shows the ico-
sadodecahedron as
composed of 6 interlocking
decagons

The icosadodecahedron
(hereinafter referred to as
i.d.) has internal planes
within it, as we have seen in
other polyhedra. The i.d. has
12 internal pentagons, 2 of
which are highlighted, one
for each pentagonal face.
And of course, the 6 'great
circle' decagons:

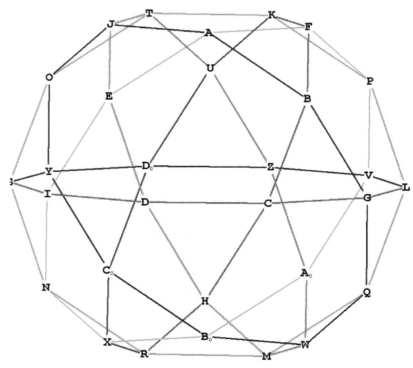

Fig. 13-2

Fig. 13-3 shows 2 of the
internal pentagons and 1
internal decagon
Fascinatingly enough, the
i.d. also has 5 hexagonal
planes! One of these is highlighted in Fig. 13-4 below.
Notice that the sides of the
hexagonal planes are, just
like the pentagonal planes,
all diagonals of the pentagonal faces. How can this be?
How can a pentagon and a
hexagon both have sides of
the same length? We'll see
later on how this happens.

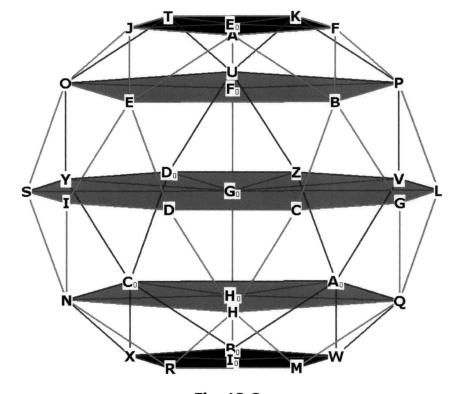

Fig. 13-3

Fig. 13-4 shows one of the 5 hexagonal planes of the icosa-dodecahedron

The i.d. is composed of equilateral triangles internally and externally. The internal equilateral triangles form from the centroid and any 2 diagonal vertices of the pentagonal faces.

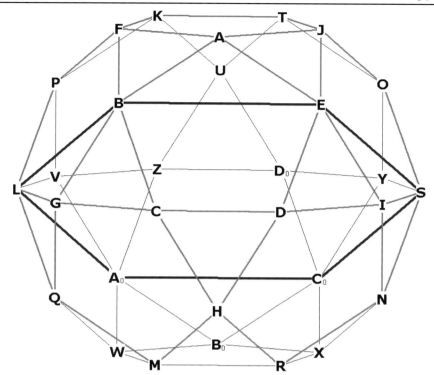

Fig. 13-4

All of these triangles (for example, triangle BGoE) are equilateral triangles, just like the 20 smaller equilateral triangles of the faces (for example, BGC). The i.d. has the property that the central angles of the diagonals of the pentagonal faces (as ∡ BGoE) are 60 degrees.

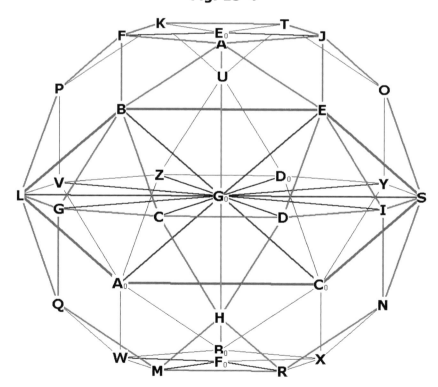

Fig. 13-5

Fig. 13-6 shows one of the hexagonal planes in relation to 2 of the pentagonal planes and one of the dodecahedron planes. Now it is clear why the hexagonal plane has the same edge length as the pentagonal planes: the hexagonal plane uses different diagonals of the pentagonal faces, and so is angled relative to the pentagonal planes. Here we show the i.d. sitting on a pentagonal face

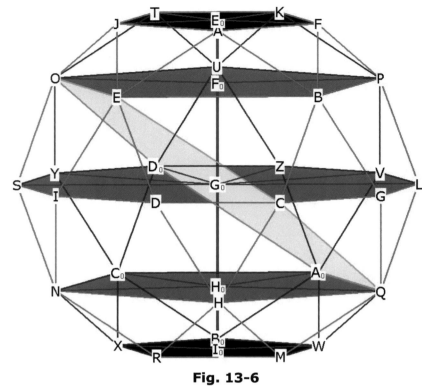

Fig. 13-6

Fig. 13-6A shows that the hexagonal plane is parallel to the top and bottom triangular faces, and also showing the two large triangular planes. Here we show the i.d. sitting on a triangular face.

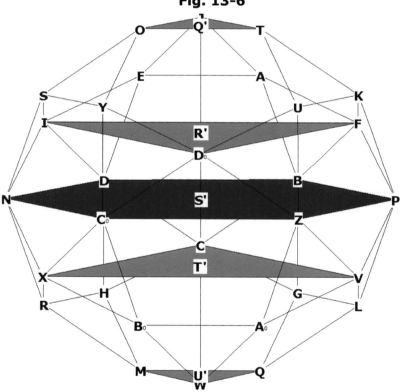

Fig. 13-6A

What is the volume of the icosadodecahedron?

We will use the pyramid method again. There are 20 triangular pyramids and 12 pentagonal pyramids.

Although the icosa-dodecahedron is a semi-regular polyhedron, all of its vertices touch on the same sphere. So the radius of the sphere surrounding the i.d. is the same for each vertex. That makes our calculations a little easier.

In order to make life simpler, it would be nice to know the relationship between the side and the radius of the i.d. I will present two ways of getting this: the absurdly simple way, and the 'brute force' method.

The simple method is to recognize that the central angle of the i.d. is 36°, and that every vertex on the i.d. is part of a decagon. Therefore any two adjacent vertices combined with the centroid forms a 36 -72 -72 isosceles triangle. We recognize this immediately as a Golden Triangle. The relationship of the short side to any of the long sides of such a triangle is therefore 1 to Φ. And so the relationship of the radius to any edge of the i.d. is r = Φ ids. (ids = icosadodecahedron side).

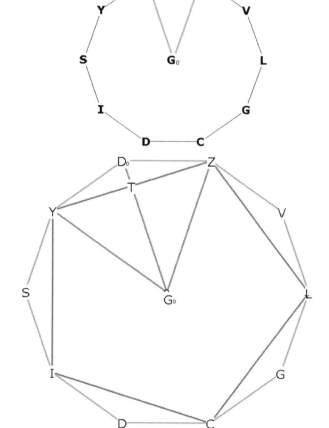

Fig 13.7 at right showing DoGo = GoZ = radius, and the golden triangle Go-Z-Do

We can calculate this relationship as follows:

Fig. 13-8 at right deriving the relationship between radius (GoZ) and side (ZDo) of the icosa dodecahedron

ZY is the side of the pentagon. Go is also the center of the pentagon.

ZT is just one half the side of the pentagon, because GoDo is an angle bisector of $\angle$ ZGoY.

We know the distance GoZ, Construction of the Pentagon Part 2 as

$\dfrac{\Phi}{\sqrt{\Phi^2 + 1}}$ side of pentagon.

GoZ is also the radius of the enclosing sphere around the i.d.

We also know from this that GoT $= \dfrac{\Phi^2}{2\sqrt{\Phi^2 + 1}}$ side of pentagon.

Therefore TDo $= \dfrac{\Phi}{\sqrt{\Phi^2 + 1}} - \dfrac{\Phi^2}{2\sqrt{\Phi^2 + 1}} = \dfrac{1}{2\Phi\sqrt{\Phi^2 + 1}}$ side of pentagon.

What is the relationship between the side of the pentagon and the side of the decagon?

Let's find ZDo, the side of the decagon, in terms of the side of the pentagon.

ZDo² = TDo² + TZ² =

$$\dfrac{1}{4\Phi^2(\Phi^2 + 1)} + 1/4 = \dfrac{1 + \Phi^2(\Phi^2 + 1)}{4\Phi^2(\Phi^2 + 1)}$$

$$= \dfrac{4\Phi^2}{4\Phi^2(\Phi^2 + 1)}$$

$$= \dfrac{1}{\Phi^2 + 1} \ * \text{ side of pentagon.}$$

ZDo $= \dfrac{1}{\sqrt{\Phi^2 + 1}}$ side of pentagon.

But r $= \dfrac{\Phi}{\sqrt{\Phi^2 + 1}}$ side of pentagon, so side of pentagon $= \dfrac{\sqrt{\Phi^2 + 1}}{\Phi}$ r .   .

Therefore, ZDo, the side of the decagon and the side of the i.d. =

$$\dfrac{1}{\sqrt{\Phi^2 + 1}} \times \dfrac{\sqrt{\Phi^2 + 1}}{\Phi} = \dfrac{1}{\Phi} r.$$

r $= \Phi \times$ ids, ids $= \dfrac{1}{\Phi}$ r. This calculation confirms the result above.

Now let's resume the volume calculation. First we'll calculate the volume of a pentagonal pyramid. Refer to Figure 9 below.

From   Area of Pentagon we know area $= \dfrac{5\Phi^2}{4\sqrt{\Phi^2 + 1}}$ ids² ,

and the distance mid-face to any vertex of a pentagon,

HoF, $= \dfrac{\Phi}{\sqrt{\Phi^2 + 1}}$ ids.

GoF $= \Phi$ *ids. GoF is just the radius of the enclosing sphere

To find the height of the pyramid, take right triangle GoHoF.

$h^2 = GoHo^2 = GoF^2 - HoF^2$

$$h^2 = \Phi^2 - \frac{\Phi^2}{\Phi^2 + 1} = \frac{\Phi^4 + \Phi^2 - \Phi^2}{\Phi^2 + 1} = \frac{\Phi^4}{\Phi^2 + 1}$$

$h = \dfrac{\Phi^2}{\sqrt{\Phi^2 + 1}}$ ids. (Note: $\dfrac{GoHo}{HoF} = \dfrac{\dfrac{\Phi^2}{\sqrt{\Phi^2 + 1}}\text{ids}}{\dfrac{\Phi}{\sqrt{\Phi^2 + 1}}\text{ids}} = \Phi$).

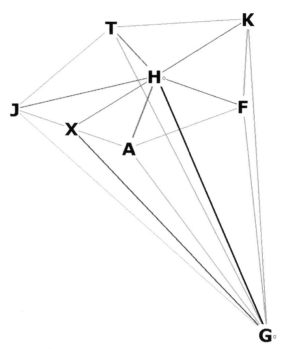

$V_{1\ \text{pent pyramid}} = 1/3 \times$ area of base $\times$ height

$$= 1/3 * \frac{5\Phi^2}{4\sqrt{\Phi^2 + 1}}\text{ids}^2 \times \frac{\Phi^2}{\sqrt{\Phi^2 + 1}}\text{ids}$$

$$= \frac{5\Phi^4}{12(\Phi^2 + 1)}\text{ids}^3$$

$$V_{12\ \text{pyramids}} = \frac{5\Phi^4}{\Phi^2 + 1}\text{ids}^3.$$

That takes care of the pentagonal pyramids. There are still 20 triangular pyramids.

Fig. 13-9 -- one of the 12 pentago-nal pyramids of the i.d.

Each triangular face area $= \dfrac{\sqrt{3}}{4}\text{ids}^2$, and the height is as follows:

Fig. 13-10 shows a triangular pyramid of the icosadodecahedron

Triangle GoZF is right by construction.

GoF $= \Phi *$ids, being the radius of the enclosing sphere.

ZF is, from Equilateral Triangle, $\dfrac{1}{\sqrt{3}}$ids.

Therefore $h^2 = GoZ^2 = GoF^2 - ZF^2 = \Phi^2 - \dfrac{1}{3} = \dfrac{3\Phi^2 - 1}{3} = \dfrac{\Phi^4}{3}$

$h = \dfrac{\Phi^2}{\sqrt{3}}$ ids $= 1.511522629$ ids.

So $V_{1\ \text{pyramid}} = 1/3$ x area of base x pyramid height =

$1/3 * \dfrac{\sqrt{3}}{4}\text{ids}^2 * \dfrac{\Phi^2}{\sqrt{3}}\text{ids}$

$= \dfrac{\Phi^2}{12}\text{ids}^3.$

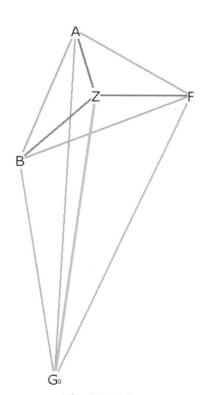

Fig.13-10

$$V_{20\ pyramids} = \frac{5\Phi^2}{3}\,ids^3.$$

$$V_{total} = V_{12\ pyramids} + V_{20\ pyramids} = \frac{5\Phi^4}{\Phi^2+1}\,ids^3 + \frac{5\Phi^2}{3}\,ids^3$$

$$= \frac{15\Phi^4 + 5\Phi^2(\Phi^2+1)}{3(\Phi^2+1)}$$

$$= \frac{20\Phi^4 + 5\Phi^2}{3(\Phi^2+1)}$$

$$= \frac{5\Phi^2(4\Phi^2+1)}{3(\Phi^2+1)}\,ids^3$$

$$= 13.83552595\ ids^3$$

This figure is slightly larger than the volume of the rhombic triacontahedron. This larger figure is easily explained by the fact that both the outer radial distance of the r.t., and the radius of the i.d., are $\Phi *$ side. Since ALL vertices of the icosadodecahedron lie on this sphere, and only 12 of the vertices of the rhombic triacontahedron do, the volume of the icosadodecahedron is naturally larger.

What is the surface area of the icosadodecahedron?

It is 12 * area of pentagonal face + 20 * area triangular face =

$$12 * \frac{5\Phi^2}{4\sqrt{\Phi^2+1}}\,ids^2 + 20(\frac{\sqrt{3}}{4})ids^2 =$$

$$\frac{15\Phi^2}{\sqrt{\Phi^2+1}}\,ids^2 + 5\sqrt{3}\,ids^2 = 29.30598285\ ids^2.$$

What is the central angle of the icosadodecahedron?

Because all of the vertices of the i.d lie along the 6 'great circle' decagons, the interior angles are all 360° / 10 = 36°. Notice that each of the central angles to adjacent vertices forms a Golden Triangle with angles 36°, 72° and 72°.

We saw earlier how the central angle between two diagonal vertices on any of the pentagonal faces, is 60° and how this forms a hexagon, and thus large internal equilateral triangles (see Fig. 13-5 and triangles BGoE, LBGo, GoES, LAoGo, AoGoCo, GoCoS). We were curious as to how the side of the hexagonal plane could be the same length as the side of the pentagonal plane. If you look at Fig. 13-6 you can see that the sides of both of these planes are a diagonal of any of the pentagonal faces. It is clear then, that like the dodecahedron and the icosahedron, the icosa dodecahedron is based on the pentagon.

Because the length of the radius (distance from centroid to any vertex) is Φ x side of i.d., and the diagonal of any pentagon is also

Φ x side of i.d., the hexagonal plane is formed.

What are the surface angles of the i.d.?

There are two of them. One is the angle forming the triangular faces, equal to 60°, and the other is the angle forming the pentagonal faces, equal to 108°.

What is the dihedral angle of the icosadodecahedron?

For this calculation, go to IcosaDodecahedron Dihedral Angle .

Let's collect some information we have already calculated:

Distance from centroid to a vertex = radius = Φ *ids.

Distance from centroid to mid-pentagonal face = $\dfrac{\Phi^2}{\sqrt{\Phi^2+1}}$ ids.

Distance from centroid to mid-triangular face = $\dfrac{\Phi^2}{\sqrt{3}}$ ids.

Distance from centroid to mid-edge = $\dfrac{\Phi\sqrt{\Phi^2+1}}{2}$ ids.

Or, in decimals,

Distance from centroid to a vertex = radius = 1.618033989 *ids.

Distance from centroid to mid-pentagonal face = 1.376381921 *ids.

Distance from centroid to mid-triangular face = 1.511522629 *ids.

Distance from centroid to mid-edge = 1.538841769 *ids.

Now let's calculate the separation of the various planes in the icosadodecahedron.

We want to get the distances between Eo, Fo, Go, Io and Ho.

Let's first get EoFo. Refer to the drawing below.

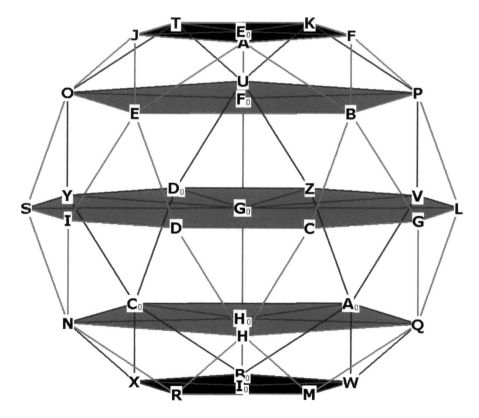

Fig. 13-3, repeated

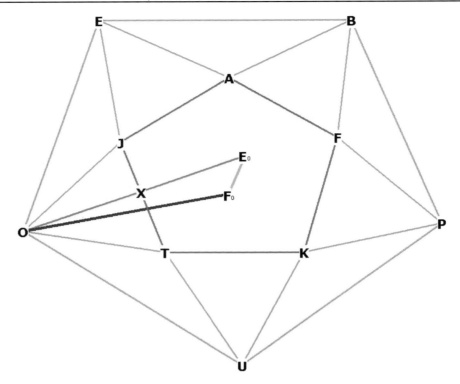

Figure 13-11a. Top view, refer to Fig. 13-3

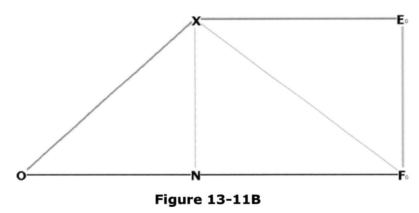

Figure 13-11B

Fig. 13-11B shows the distance between the first 2 pentagonal planes ($E_o F_o$) of the icosado-decahedron. Note that although OXE_o appears to be a straight line in Fig. 13-11a, it is actually the dihedral angle of the solid.

The top pentagonal plane on the i.d. is the plane AJTFK, in purple in Fig. 13-11a and the top plane in Fig. 13-3. The large pentagonal plane UPBEO is marked in orange.

The point E_o is the center of the small pentagon, F_o is the center of the large pentagon.

AJTFK is on top of UPBEO.

The angle $F_o E_o X$ is right. $E_o F_o$ is the line perpendicular to the plane AJTFK from the center of the plane UPBEO, and is part of the diameter of the sphere which runs through the centroid O.

The angle $OF_o E_o$ is also right, the line $E_o F_o$ is perpendicular to the plane UPBEO.

OX and XE_o can be easily calculated.

OX is just the height of an equilateral triangle, $\frac{\sqrt{3}}{2}$ ids , and XEo is the distance, in a pentagon, from the center to any mid-edge. We know from Construction of the Pentagon Part 2 that this distance is $\frac{\Phi^2}{2\sqrt{\Phi^2+1}}$ ids.

The distance OFo is just the distance from the center of the large pentagon, to one of its vertexes. From Construction of the Pentagon Part 2 we know this distance to be

$\frac{\Phi}{\sqrt{\Phi^2+1}}$ * side of the large pentagon.

We may draw a perpendicular bisector from X to N on the line OFo.

$\angle$ EoXN and $\angle$ XNFo are both right.

We now have a rectangle XNFoEo with 4 right angles at the corners, and XN = EoFo, XEo = NFo.

Now we can find the distance ON, for it is just OFo - NFo.

Then we can work with triangle ONX to find XN and we have EoFO.

First we must find OFo in terms of the side of the icosadodecahedron.

Remember that OFo is $\frac{\Phi}{\sqrt{\Phi^2+1}}$ * side of the large pentagon. But the side of the large pentagon

is just the diagonal of any the pentagonal faces of the i.d.

Therefore the side of the large pentagon is Φ * ids.

We can then write OFo = $\frac{\Phi}{\sqrt{\Phi^2+1}} \times \Phi$ ids = $\frac{\Phi^2}{\sqrt{\Phi^2+1}}$ ids.

(Note that this distance is precisely the distance from the centroid to the mid-face of any of the pentagonal faces!)

NFo = XEo = $\frac{\Phi^2}{2\sqrt{\Phi^2+1}}$ ids.

Then ON = OFo - NFo $= \frac{\Phi^2}{\sqrt{\Phi^2+1}}$ ids $- \frac{\Phi^2}{2\sqrt{\Phi^2+1}}$ ids

$= \frac{\Phi^2}{2\sqrt{\Phi^2+1}}$ ids.

From this calculation we gather that ON = NFo.

XN² = OX² - ON² =

$\frac{3}{4} - \frac{\Phi^4}{4(\Phi^2+1)} = \frac{3(\Phi^2+1)-\Phi^4}{4(\Phi^2+1)} = \frac{4}{4(\Phi^2+1)} = \frac{1}{\Phi^2+1}$.

XN = EoFo $= \frac{1}{\sqrt{\Phi^2+1}}$ ids.

Therefore the distance between the first two pentagonal planes is $\dfrac{1}{\sqrt{\Phi^2+1}}$ ids.

Now we can easily find FoGo, the distance between the large pentagonal plane and the plane of the decagon which contains the centroid.

This distance is just EoGo - EoFo.

EoGo is the distance from the centroid to mid-face of pentagon,

which we found above to be $\dfrac{\Phi^2}{\sqrt{\Phi^2+1}}$ ids.

So FoGo $= \dfrac{\Phi^2}{\sqrt{\Phi^2+1}}$ ids $- \dfrac{1}{\sqrt{\Phi^2+1}}$ ids $= \dfrac{\Phi^2-1}{\sqrt{\Phi^2+1}}$ ids

$\qquad = \dfrac{\Phi}{\sqrt{\Phi^2+1}}$ ids.

What is the relationship between EoFo, FoGo and EoGo?

EoGo / FoGo $= \dfrac{\dfrac{\Phi^2}{\sqrt{\Phi^2+1}}}{\dfrac{\Phi}{\sqrt{\Phi^2+1}}} = \dfrac{\Phi^2}{\Phi} = \Phi!$

Therefore the radius EoGo is divided in Mean and Extreme Ratio at Fo.

Table of relationships of the distances between the pentagonal planes

If we let EoFo = 1, then

Eo_____	_____
1	
Fo_____	
	Φ^2
Φ	
Go_____	_____
Φ	
	Φ^2
Ho_____	
1	
Io_____	_____

What are the distances between the triangular planes and the hexagonal plane which runs through the centroid?   (Refer back to Fig. 13-6A above).

We want R'Q' and T'U'. The karge equilateral triangle DoIF in Fig. 13-6A has sides equal to the distance DoF = FI = DoI in Fig. 13-12:

Fig. 13-12 shows the length of the sides of the large triangular planes. You can see this in Fig. 13-6A by examining the vertices Do, Y, S, I which are part of one of the 'great circle' decagons. The length of each side of the triangular plane is DoI. The other two legs of this large equilateral triangle are each part of a different decagon.

From *Decagon* (appendix D) we know the edge length of each side of tri-angle DoIF is Φ^2 ids. In Appendix D, this will be seen as the distance DA (see p. 145)

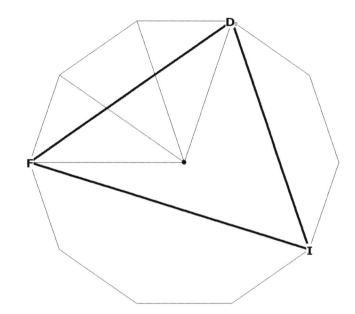

Fig.13-12

R' is the center of the triangular face DoIF. T' is the center of the triangular face CXV, and O' is the centroid. We want to find R'T', the distance between the two triangular planes. We will first find O'R'.

The solution is quite simple, actually. The distance DoR' is known. DoR' is the dis-tance from the center of an equilateral triangle to an i.d. vertex. O'Do is also known. O'Do is just the distance from the centroid to a vertex of the i.d. R' lies directly above O'. Therefore the triangle O'R'Do is right and we may write

$$\overrightarrow{O'R'}^2 = \overrightarrow{O'D_0}^2 - \overrightarrow{R'D_0}^2$$

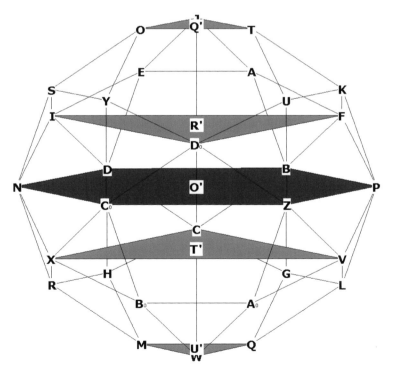

Fig.13-6A, repeated

O'Do = Φ x ids, and from Equilateral Triangle we know that R'Do =

$\dfrac{1}{\sqrt{3}} \times$ side of triangle, in this case, Φ^2 ids .

$$\overline{O'R'}^{\,2} = \Phi^2 \; ids^2 - \left(\dfrac{\Phi^2}{\sqrt{3}}\right)^2 \; ids^2 = \dfrac{3\Phi^2 - \Phi^4}{3} ids^2 = \dfrac{3\Phi^2 - 3\Phi^2 + 1}{3} = \dfrac{1}{3} ids^2.$$

$$O'R' = \dfrac{1}{\sqrt{3}} ids = 0.577350269 \; ids.$$

Because the analysis is precisely the same for O'T', the distance O'T' also equals $\dfrac{1}{\sqrt{3}} ids$, and

$$R'T' = \dfrac{2}{\sqrt{3}} ids = 1.154700538 \; ids.$$

What is the distance Q'R'? This is the distance between the triangular face OJT and the plane DoIF. The analysis is the same as above, except that the side of the triangular plane OJT is now

ids, instead of $\Phi^2 \times$ ids (Why? Because OJT is a face of the i.d.) Therefore

O'R' is the distance from the centroid to mid-triangular face (p. 103)

Q'R' = O'Q' − O'R'

$$Q'R' = \dfrac{\Phi^2}{\sqrt{3}} ids - \dfrac{1}{\sqrt{3}} ids$$

$$= \dfrac{\Phi^2 - 1}{\sqrt{3}} ids = \dfrac{\Phi}{\sqrt{3}} ids$$

$$= 0.934172359 \; ids$$

$$\dfrac{Q'R'}{O'R'} = \dfrac{\left(\dfrac{\Phi}{\sqrt{3}}\right)}{\left(\dfrac{1}{\sqrt{3}}\right)} = \Phi$$

What is the angle any of the hexagonal planes make with the plane of the decagon?

In Fig. 13-6A, the hexagonal plane is highlighted in purple.

This can be seen by observing the plane of the hexagon as it intersects the plane of the decagon in Fig. 13-6.

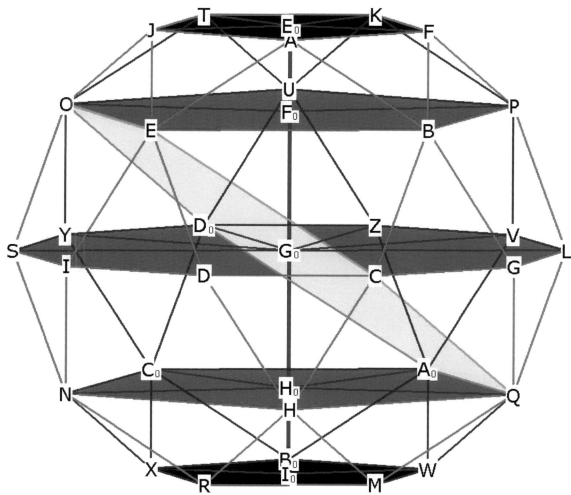

Fig.13-6, repeated

Notice that $\angle$ OGoS is the angle of the hexagonal plane as it intersects with the central decagon, and that is it also a central angle of the i.d. This angle, of course, is 360 / 10 = 36°.

Conclusions:

What have we learned about the icosadodecahedron?

Mainly, that it is pentagonally based.

The triangular faces fill in the gaps between the pentagonal faces.

All of the radii form Golden Triangles with any 2 adjacent vertices.

The spacing of the pentagonal and decagonal internal planes are based upon the division of the space into Mean and Extreme Ratio.

The Dihedral Angle of the Icosa–Dodecahedron

Consider Fig. 13-13:. We see that the dihedral angle is ∡LXA, being the intersection of the 2 planes BGLPF and FBA. I have drawn the line LXA which goes directly through the middle of both planes and forms a 'roof' over them.

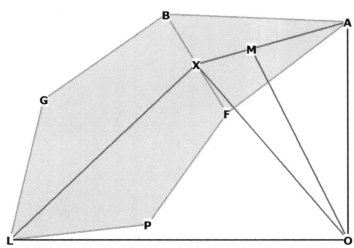

Fig. 13-13 The dihedral angle of the icosa–dodecahedron

If you build an icosa–dodecahedron you will see immediately that the angle from A to the centroid O and back to L, or ∡LOA, is right.

LO = AO = radius of enclosing sphere = Φ * ids.

Our plan of attack for getting ∡LXA will be as follows:

Note that ZX is drawn parallel to OA, and thus also perpendicular to LO. Therefore ∡XZO and ∡XZA are right.

AM and MX are the distances, respectively, from the vertex of an equilateral triangle to the mid-face, and from the mid-face to a mid-edge.

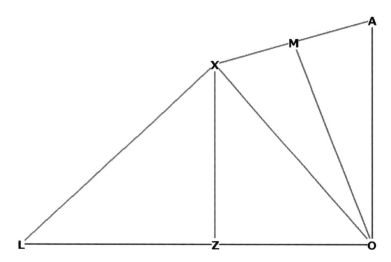

Fig. 13-14: dihedral angle, revisited

From Equilateral Triangle, we know these to be $\dfrac{1}{\sqrt{3}}$ ids and $\dfrac{1}{2\sqrt{3}}$ ids.

XL is also known, being the height of a pentagon.

From Pentagon Construction Part 2 we know this distance to be $\dfrac{\Phi\sqrt{\Phi^2+1}}{2}$ ids.

In Fig. 13-14, the line OM is drawn perpendicular to the triangular face at its midpoint. Therefore the angle OMA and the angle OMX are right.

Note that from the above data, we can conclude that the triangles OMA, OMX, XZO and XZL are right.

OM can be determined because we know OA and AM.

Therefore the angle MOA can be calculated.

With OM and MX known, OX can be determined, because triangle OMX is right.

Therefore the angle MOX can be calculated.

Now $\angle$ AOX is known, and we can get $\angle$ XOZ, because we know $\angle$ LOA is right.

Because triangle XZO is right, and we know $\angle$ XOZ, the angle ZXO can be calculated.

Now we know the angle ZXA, it being the sum of $\angle$ ZXO and $\angle$ OXA.

Now we are left with triangle LZX, which is right by construction.

XZ can be determined from the data in triangle XZO, and then $\angle$ ZXL can be calculated from triangle LZX.

The dihedral angle LXA is then the sum of $\angle$ ZXL and $\angle$ ZXA.

That was quite a lot of work, but now the fun stuff is done, which for me is figuring out the geometry. Now it's just a bunch of calculations!

Our work is made easier because we can get the distances OM and OX from our previous calculations in Icosa–Dodecahedron. OM is just the distance from the centroid of the i.d. to a mid–triangular face, and OX is just the distance from the centroid to any mid–edge.

We determined OM = $\dfrac{\Phi^2}{\sqrt{3}}$ ids and OX = $\dfrac{\Phi\sqrt{\Phi^2 + 1}}{2}$ ids.

Let's get going!

$$\sin(\angle MOA) = AM / OA = \dfrac{\dfrac{1}{\sqrt{3}}}{\Phi} = \dfrac{1}{\sqrt{3}\Phi}.$$

$\angle$ MOA = 20.90515744°.

$$\tan(\angle MOX) = MX / OM = \dfrac{\dfrac{1}{2\sqrt{3}}}{\dfrac{\Phi^2}{\sqrt{3}}} = \dfrac{1}{2\Phi^2}.$$

$\angle$ MOX = 10.81231696°.

$\angle$ XOA is therefore $\angle$ MOA + $\angle$ MOX = 31.71747441°.

Well, well well. Here we recognize an angle which is a part of our good friend the Phi Right Triangle with sides divided in Mean and Extreme Ratio.

Triangle XOA is not right, but triangle XZO is.

Since $\angle$ LOA is right,

$\angle$ LOX = $\angle$ ZOX = $\angle$ LOA - $\angle$ XOA = 90 - 31.71747441 = 58.28252559°.

Therefore $\angle$ ZXO = 31.71747441° and triangle XZO has sides divided in Mean and Extreme Ratio.

Therefore we know XZ / OZ = Ø.

Now

$$\tan(\angle OXM) = OM / MX = \frac{\left(\dfrac{\Phi^2}{\sqrt{3}}\right)}{\left(\dfrac{1}{2\sqrt{3}}\right)} = 2\Phi^2.$$

$\angle OXM = 79.18768304°$. (Note: We could also have calculated $\angle OXM$ using the following reasoning: Because triangle OMX is right,

$\angle OXM = \angle OXA = 180 - 90 - 10.81231696° = 79.18768304°$).

So $\angle ZXA = \angle ZXO + \angle OXM = 110.9051574°$.

Or more precisely, $\angle ZXA = \arctan\left(\dfrac{1}{\Phi}\right) + \arctan\left(2\Phi^2\right)$.

We are almost there.

If we can find $\angle ZXL$, we can finally determine $\angle LXA$.

We know XL. We need to get either LZ or XZ.

Lets get LZ.

LZ = LO - ZO. What is ZO?

To determine that, we need OX, which we already know.

Now we can write:

$\cos(\angle ZOX) = ZO / OX, \quad ZO = OX * \cos(\angle ZOX)$.

We know that triangle XZO is a Φ triangle and therefore

$$\cos(\angle ZOX) = \frac{1}{\sqrt{\Phi^2 + 1}}.$$

So $ZO = \dfrac{\Phi\sqrt{\Phi^2 + 1}}{2} \times \dfrac{1}{\sqrt{\Phi^2 + 1}} = \dfrac{\Phi}{2} = 0.809016995$.

Now since LO = radius = $\Phi \times$ ids, then LZ = LO - ZO = $\Phi - \dfrac{\Phi}{2} = \dfrac{\Phi}{2}$ ids.

Therefore LZ = ZO.

$$\sin(\angle ZXL) = LZ / XL = \frac{\dfrac{\Phi}{2}}{\dfrac{\Phi\sqrt{\Phi^2 + 1}}{2}} = \frac{1}{\sqrt{\Phi^2 + 1}}.$$

$\angle ZXL = 31.71747441°$.

So the dihedral angle LXA = $\angle ZXL + \angle ZXA$ =

$110.9051574° + 31.71747441° = \mathbf{142.6226318°}$.

Note that OX = XL and so triangle XZO congruent to triangle XZL.

Triangle XLO is isosceles.

XL / LZ also equals Φ.

Icosa Dodecahedron Reference Chart

Volume (edge)	Volume in Unit Sphere	Surface Area (edge)	Surface Area in Unit Sphere
13.83552595 s³	3.266124627 r³	29.30598285 s²	11.19388937 r²

Central Angle:	Dihedral Angle:	Surface Angles:
		60°
36°	142.6226318°	108°

Centroid To: Vertex	Centroid To: Mid–Edge	Centroid To: Mid–Triangular Face
1.0 r	0.951056516 r	0.934172359 r
1.618033989 s	1.538841769 s	1.511522629 s

Centroid To:	Side / radius
Mid- Pentagonal Face	0.618033989
1.376381921 s	
0.850650809 r	

Chapter 14 – The Rhombic Triacontahedron

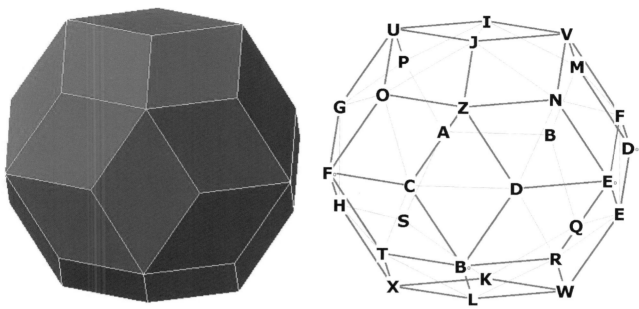

Fig. 14-1 -- showing the front portion of the Rhombic Triacontahedron (red) with the sides of the dodecahedron (gray)

The Rhombic Triacontahedron is an extremely fascinating polyhedron. It is built around the dodecahedron, and like the dodecahedron, it has many phi relationships within it. Remarkably, this polyhedron contains all five of the Platonic Solids directly on its vertices, and shows the proper relationship between them.

The Rhombic Triacontahedron (hereinafter referred to as r.t.) has 30 faces, 60 edges, and 32 vertices.

Notice pentagon CDNJO, then notice Z. Z is raised off the pentagon (a distance we shall find later on) and Z is connected to all 5 vertices of the pentagonal face of the dodecahedron. This can be seen even more clearly at pentagon GPIJO, where U rises off its face. The pentagon GPIJO has all of its vertices connected to U. Of course, we could have drawn this figure without the dodecahedron faces, which is obviously not part of the r.t., but it helps for clarity.

Notice that the faces of the r.t. are diamond-shaped, somewhat like the rhombic dodecahedron, but these faces are longer and skinnier. Look at the r.t. face UOZJ at the upper left. U and Z are the long-axis vertices. Notice that the line between O and J, the short-axis vertices, form one of the edges (sides) of the dodecahedron.

The r.t. is more clearly spherical than any of the polyhedra we have studied so far.

In Fig. 14-2 below, we see that the rhombic triacontahedron has internal pentagonal planes just like the icosahedron and the dodecahedron!

Note the large highlighted pentagons. The lengths of all of the large internal pentagon sides are precisely the long axis of every r.t. face! For example, look at VZ and ZFo in the top pentagonal plane and EoBo in the lower pentagonal plane.

The r.t. has a circumsphere and an inner sphere. The circumsphere goes around all 12 of the vertices that rise off the 12 faces of the dodecahedron.

The diameter of the circumsphere is UW. UW is in green in Fig. 14-2. O' is the centroid.

The inner sphere touches all of the 20 vertices of the dodecahedron. The diameter of the inner sphere would be, for example, IL. UW > IL.

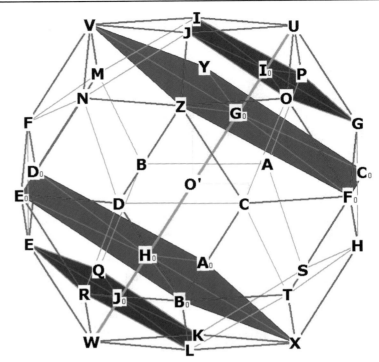

Fig. 14-2 -- showing the internal pentagonal planes of the r.t. Figure 2 is rotated 180 degrees with respect to Fig. 14-1

This leads us to think that possibly, the relationship between U, Go,O',Ho and W will be similar to those of the dodecahedron and the icosahedron. Later on, we will see that this is indeed the case. Notice also that the rhombic triacontahedron contains on its vertices not only a dodecahedron, but an icosahedron as well! Notice that the 12 vertices that rise off the dodecahedron faces provide the 12 vertices of the icosahedron. Remember that the dual of the dodecahedron is the icosahedron and that the dual is formed by taking points at the center of the faces.

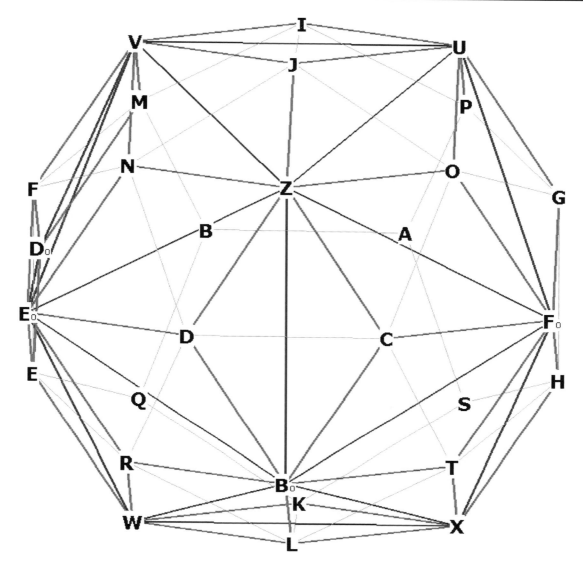

**Fig. 14-3 showing the 12 vertices of the icosahedron (blue) within the rhombic triac-
ontahedron**

The sides of the icosahedron are the long axes of the r.t. faces.

The sides of the dodecahedron are the short axes of the r.t. faces.

Therefore the rhombic triacontahedron is just the combination of the dodecahedron with its
dual, the icosahedron.

The rhombic triacontahedron is the model nature uses to demonstrate the true relationship between the side of the icosahedron and the side of the dodecahedron. We will see later on that
these relationships are based on Φ.

Note also that when a cube is inscribed within 8 of the 12 vertices belonging to the dodecahedron within the rhombic triacontahedron, the lengths of the cube sides are equal to the long axis
of any of the rhombic triacontahedron rhombi. Furthermore, the sides of the cube are equal to
any of the diagonals of the pentagonal faces of the dodecahedron:

The sides of the cube, in blue, are also the diagonals of the pentagonal faces of the dodecahedron within the rhombic triacontahedron. Since the tetrahedron (one tetrahedron in green, the other in purple) and the octahedron (in orange) can be inscribed within the cube, the rhombic triacontahedron shows the precise relationship between the Platonic Solids! See Fig. 14-3B below.

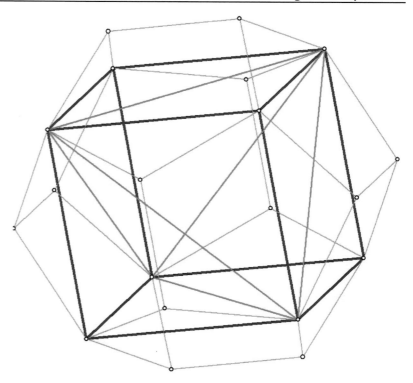

Fig.14-3A

The Rhombic Triacontahedron therefore elegantly describes the nesting of the five Platonic Solids: icosahedron, dodecahedron, cube, tetrahedron, octahedron. When the sides of the octahedron are divided in Mean and Extreme (Phi) Ratio, another icosahedron is formed. This begins the process all over again, and shows that the 5 nested Platonic Solids may not only grow and contract to infinity, but do so in a perfectly harmonious way.

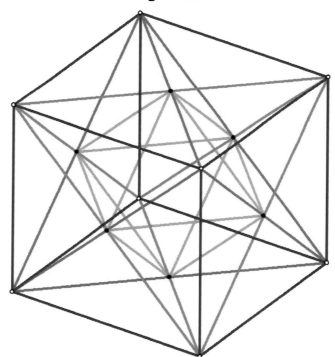

Fig. 14-3B: The cube, 2 interlocking tetrahedrons, and the octahedron inside a cube

Interestingly, the icosahedron is formed within the octahedron by dividing each edge of the octahedron in Phi Ratio. Here again is an elegant link between root 2 and root 3 geometry, and Phi.

What is the volume of the rhombic triacontahedron?

As before we use the pyramid method. There are 30 faces, so there are 30 pyramids. Imagine a point at the very center of the r.t. If you connect that point up with one of the faces, you will have a pyramid that looks like Fig. 14-4.

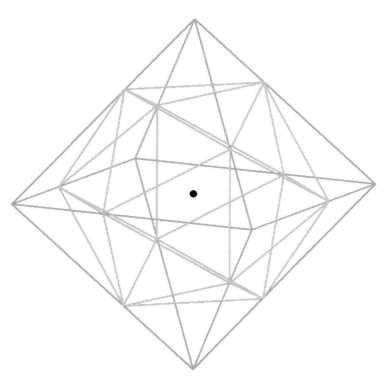

Fig. 14-3C: Showing how the icosahedron nests within the octahedron.

O'Z is the height of the pyramid. O'U = O'V = radius of outer sphere which touches the 12 r.t. vertices which are raised off the center of the dodecahedron face. O'I = O'J = radius of inner sphere which touches all 20 of the short-axis r.t. vertices, which are also the vertices of the dodecahedron
We want to find the volume of the r.t. in terms of the r.t. side. However, we don't know the length of the r.t side! But we do know the length of the side of the dodecahedron, in terms of a unit sphere which touches all the vertices of the dodecahedron. So let's try to get the r.t. side (hereinafter referred to as rts) in terms of the dodecahedron side.

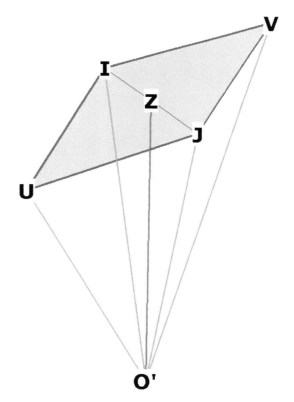

Fig. 14-4: one of the 30 pyramids of the rhombic triacontahedron.

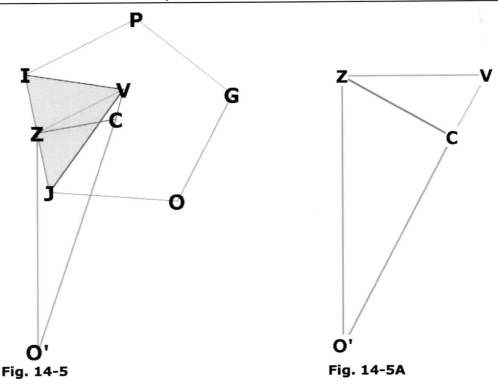

Fig. 14-5   **Fig. 14-5A**

Fig. 14-5 and 14-5A show the following important data:

V is one of the 12 raised vertices of the r.t. off the face of the dodecahedron.

C is the center of the dodecahedron face. C lies in the plane of IJOGP, and is directly below V.

Z is the center of the r.t. face IUJV (see Fig. 14-4), and also the mid-edge of the side of the do-dec.

IVJ (in red) is one-half of the r.t. diamond face IUJV (see Fig. 14-4)

O'Z is the distance from centroid to mid-edge of the dodec face. This is the height of the r.t. pyramid.

ZV is the distance from mid-edge of dodecahedron to the raised vertex V off the dodecahedron face.

ZC is the distance from mid-edge of dodecahedron to center of dodecahedron face.

CV is the distance of V off the center of the dodecahedron face.

Figure 5A shows that O'CV is a straight line, so that

∡ ZO'C = ∡ ZO'V.

∡ O'ZV is right.

∡ O'CZ is right.

From this data we can show that the triangles O'ZV, O'CZ, and CZV are similar by angle-side-angle.

First we show the triangle O'ZV and O'CZ are similar:

∡ O'ZV is right, so is O'CZ.

O'Z is common to both triangles.

$\angle$ ZO'V = $\angle$ ZO'C.   Therefore both triangles are similar by ASA.

Now we show that triangle CZV is similar to triangle O'ZV by angle-side-angle.

$\angle$ ZCV and $\angle$ O'ZV are right.

ZV is common to both triangles.

$\angle$ O'VZ = $\angle$ CVZ.

Therefore both triangles are similar by angle-side-angle.

With this information we can determine CV, the distance of the vertex V off of the dodec face, and ZV, which will enable us to get the side of the r.t. in terms of the side of the dodecahedron.

Since all 3 triangles are similar, we can write the following relationship:

CZ / O'C = CV / CZ.

The distances O'Z, O'C, and CZ are known. From Dodecahedron it is known that

$$CZ = \frac{\Phi^2}{2\sqrt{\Phi^2 + 1}} ds. \quad \text{(ds means dodecahedron side)}$$

$$O'C = \frac{\Phi^3}{2\sqrt{\Phi^2 + 1}} ds.$$

$$O'Z = \frac{\Phi^2}{2} ds = h, \text{ the height of the r.t. pyramid.}$$

$$\text{So CV / CZ} = \frac{\left(\dfrac{\Phi^2 s}{2\sqrt{\Phi^2 + 1}}\right)}{\left(\dfrac{\Phi^3 s}{2\sqrt{\Phi^2 + 1}}\right)} = \frac{1}{\Phi}!$$

$$CV = \frac{1}{\Phi} * CZ = \frac{\Phi}{2\sqrt{\Phi^2 + 1}} ds = 0.425325404 \ ds.$$

Now we have CV, the distance from the plane of the dodecahedron to the rhombic triacontahedron "cap" over the dodecahedron face, in terms of the dodecahedron side.

Now we need to find ZV, so that we can get IV, the length of the side or edge of the rhombic triacontahedron. We can write the following relationship:

$$\frac{CZ}{ZV} = \frac{O'C}{O'Z} = \frac{\dfrac{\Phi^3}{2\sqrt{\Phi^2 + 1}}}{\dfrac{\Phi^2}{2}} = \frac{\Phi}{\sqrt{\Phi^2 + 1}} ds.$$

$$ZV = \frac{\sqrt{\Phi^2 + 1}}{\Phi} * CZ = \frac{\sqrt{\Phi^2 + 1}}{\Phi} * \frac{\Phi^2}{2\sqrt{\Phi^2 + 1}} = \frac{\Phi}{2} ds = 0.809016995 \ ds.$$

Refer back to Figs. 14-4 and 14-5. Now that we have ZV, we can find the side of the r.t., IV. The triangle IVJ in Fig. 14-4 and 14-5 is one-half of an r.t face. We have already calculated ZV, and we know that IZ is just one–half the side of the side of the dodecahedron, ds. We also know that the angle IZV is right by construction.

Therefore, by the Pythagorean Theorem,

$$\overline{IV}^2 = \overline{IZ}^2 + \overline{ZV}^2 = \frac{1}{4} + \frac{\Phi^2}{4} = \frac{\Phi^2+1}{4}\,ds^2.$$

$$IV = rts = \frac{\sqrt{\Phi^2+1}}{2}\,ds, \quad \text{and } ds = \frac{2}{\sqrt{\Phi^2+1}}\,rts = 1.0514562224 \text{ rts.}$$

Here we have established an important fact: we have related the side of the rhombic triaconta-hedron, or rts, to the side of the dodecahedron and, therefore, to the radius of the unit sphere which encloses all 20 vertices of the dodecahedron and, in turn, the short axis vertices of the rhombic triacontahedron.

We can now describe the distance of any vertex of the r.t. off the plane of the dodecahedron, CV, in terms of the side of the r.t. We may now write

$$CV = \frac{\Phi}{2\sqrt{\Phi^2+1}}\,ds = \frac{\Phi}{2\sqrt{\Phi^2+1}} * \frac{2}{\sqrt{\Phi^2+1}}\,rts = \frac{\Phi}{\Phi^2+1}\,rts$$

$$CV = 0.447213596 \text{ rts.}$$

Let's now find the radius of the outer sphere of the r.t. in terms of the side of the r.t. itself. Re-member that the outer sphere touches all 12 long-axis vertices of the r.t., and that these 12 vertices are the vertices of an icosahedron

$r_{outer} = O'V = O'C + CV.$ We need to convert these distances so that they are related to rts, not ds.

$$O'C = \frac{\Phi^3}{2\sqrt{\Phi^2+1}}\,ds = \frac{\Phi^3}{2\sqrt{\Phi^2+1}} * \frac{2}{\sqrt{\Phi^2+1}}\,rts = \frac{\Phi^3}{\Phi^2+1}\,rts.$$

$$CV = \frac{\Phi}{2\sqrt{\Phi^2+1}}\,ds = \frac{\Phi}{2\sqrt{\Phi^2+1}} * \frac{2}{\sqrt{\Phi^2+1}}\,rts = \frac{\Phi}{\Phi^2+1}\,rts.$$

$$\text{So } r_{outer} = O'V = \frac{\Phi^3}{\Phi^2+1}\,rts + \frac{\Phi}{\Phi^2+1}\,rts = \frac{\Phi(\Phi^2+1)}{\Phi^2+1}\,rts = \Phi.$$

Therefore $r_{outer} = O'V = \Phi * rts. \quad rts = \dfrac{r_{outer}}{\Phi}.$

In the Rhombic Triacontahedron, the relationship between the side and the radius of the enclos-ing sphere is Phi. Interestingly, this is precisely what we found with the icosa–dodecahedron, the dual of the rhombic triacontahedron. Great minds think alike, as they say.

(BTW, what did the rhombic triacontahedron say to the icosadodecahedron? "You're Phi–ntastic." OK, really bad joke, but I couldn't resist!)

What is r_{inner}? This is just the unit sphere which touches all 20 vertices of the dodecahedron. We

know from Dodecahedron that $r_{inner} = \dfrac{\sqrt{3}\Phi}{2}\,ds.$

Converting this to the side of the r.t. we have:

$$r_{inner} = \frac{\sqrt{3}\Phi}{2}\,ds = \frac{\sqrt{3}\Phi}{2} * \frac{2}{\sqrt{\Phi^2+1}}\,rts = \frac{\sqrt{3}\Phi}{\sqrt{\Phi^2+1}}\,rts = 1.47337042 \text{ rts.}$$

Note how the $\sqrt{3}$ appears in the numerator. The dodecahedron is the link between the $\sqrt{3}$ geometry of the cube, tetrahedron and octahedron, and Φ, which appears over and over again in more complex polyhedra.

Now we have enough information to calculate the volume of the rhombic triacontahedron in terms of it's own side. Refer back to Figures 4 and 5 for diagrams.

We can calculate the area of the r.t. face, because we know IJ and ZV, which we can use to give us the area of one-half the face of the r.t. We see from Figure 4 that O'Z, the height of the r.t. pyramid, is just the distance from the centroid to the mid-edge of any of the sides of the dodecahedron. We know from Dodecahedron, that this distance is $\dfrac{\Phi^2}{2}$ ds . We have to do some conversions of these values first, to get them all in terms of the rts.

To get the area of the r.t. face, divide it into 2 identical triangles at the short axis IJ, find the area of one triangle, and multiply by 2..

The area of any triangle is 1/2 * base * height, so the area of the r.t. face is twice this value.

IJ is just the side of the dodecahedron, and it is the base of our triangle.

$$IJ = ds = \frac{2}{\sqrt{\Phi^2 + 1}} \, rts.$$

$$ZV = \text{height of triangle} = \frac{\Phi}{2} ds = \frac{\Phi}{2} \times \frac{2}{\sqrt{\Phi^2 + 1}} \, rts = \frac{\Phi}{\sqrt{\Phi^2 + 1}} \, rts.$$

$$\text{area of 1 triangle} = \frac{1}{2} \times \frac{2}{\sqrt{\Phi^2 + 1}} \times \frac{\Phi}{\sqrt{\Phi^2 + 1}} = \frac{\Phi}{\Phi^2 + 1} \, rts^2.$$

$$\text{Area of 1 diamond r.t. face} = \frac{2\Phi}{\Phi^2 + 1} \, rts^2.$$

$$Volume_{1 \, r.t. \, pyramid} = \frac{1}{3} \times \text{ area of face } \times \text{ pyramid height}$$

$$= \frac{1}{3} \times \text{ area of face } \times O'Z$$

$$= \frac{1}{3} \times \frac{2\Phi}{\Phi^2 + 1} rts^2 \times \frac{\Phi^2}{2} ds$$

$$= \frac{1}{3} \times \frac{2\Phi}{\Phi^2 + 1} rts^2 \times \frac{\Phi^2}{2} \times \frac{2}{\sqrt{\Phi^2 + 1}} rts$$

$$= \frac{1}{3} \times \frac{2\Phi}{\Phi^2 + 1} rts^2 \times \frac{\Phi^2}{\sqrt{\Phi^2 + 1}} rts$$

$$= \frac{2\Phi^3}{3(\Phi^2 + 1)^{3/2}} rts^3$$

There are 30 pyramids for 30 faces so total volume V_{total} is:

$$V_{total} = 30 * \frac{2\Phi^3}{3(\Phi^2 + 1)^{3/2}} rts^3 = \frac{20\Phi^3}{(\Phi^2 + 1)^{3/2}} rts^3 = 12.31073415 \, rts^3.$$

The volume of the r.t. can be calculated another way. I include this calculation not because it's necessary, but to show that there is more than one way to get the answer. Math and geometry is more about the ability to think with the material than a series of boring "plug–in–the–numbers" exercises.

Since the r.t. is built upon the dodecahedron, the r.t. volume is just the volume of the dodecahedron + the extra volume of all of the little 12 pentagonal pyramids formed from the raised vertices off the 12 pentagonal faces of the dodecahedron. To see this, check Figure 1 again and look at U-IJOGP or V- IJNFM.

The volume of the dodecahedron is, from Dodecahedron , $\dfrac{5\Phi^5}{2(\Phi^2+1)}\,ds^3$.

Converting this to the side of the r.t. we get:

$$V_{dodecahedron} = \frac{5\Phi^5}{2(\Phi^2+1)} \times \left(\frac{2}{\sqrt{(\Phi^2+1)}}\right)^3 rts^3$$

$$= \frac{5\Phi^5}{2(\Phi^2+1)} \times \frac{8}{(\Phi^2+1)^{\frac{3}{2}}} rts^3$$

$$= \frac{20\Phi^5}{(\Phi^2+1)^{\frac{5}{2}}} rts^3 = 8.908130915\ rts^3.$$

The volume of each of the 12 "extra" pyramids =

1/3 * area of pentagon * CV (height of each raised vertex off of the face of the dodecahedron)

$$= \frac{1}{3} * \frac{5\Phi^2}{4\sqrt{\Phi^2+1}}\,ds^2 * \frac{\Phi}{\Phi^2+1}\,rts.$$

We only have one problem: one of our values is in terms of the dodecahedron side. We need to convert that to the side of the r.t.

$$\frac{5\Phi^2}{4\sqrt{\Phi^2+1}}\,ds^2 = \frac{5\Phi^2}{4\sqrt{\Phi^2+1}} * \left(\frac{2}{\sqrt{\Phi^2+1}}\right)^2 rts^2$$

$$= \frac{5\Phi^2}{4\sqrt{\Phi^2+1}} * \frac{4}{\Phi^2+1}\,rts^2$$

$$= \frac{5\Phi^2}{(\Phi^2+1)^{3/2}}\,rts^2.$$

$$V_{1\ 'extra'\ pyramid} = \frac{1}{3} * \frac{5\Phi^2}{(\Phi^2+1)^{3/2}}\,rts^2 * \frac{\Phi}{\Phi^2+1}\,rts$$

$$= \frac{5\Phi^3}{3(\Phi^2+1)^{5/2}}\,rts^3.$$

$$V_{12\ 'extra'\ pyramids} = \frac{20\Phi^3}{(\Phi^2+1)^{5/2}}\,rts^3.$$

$$V_{r.t.} = V_{dodecahedron} + V_{'extra'} = \frac{20\Phi^5 + 20\Phi^3}{(\Phi^2+1)^{5/2}}\,rts^3 = 12.31073415...\ rts^3.$$

What is the surface area of the rhombic triacontahedron?

It is just $30 \times$ area of each face $= 30 \times \dfrac{2\Phi}{\Phi^2+1}\,rts^2 = \dfrac{60\Phi}{\Phi^2+1}\,rts^2 = 26.83281573\ rts^2.$

Before we calculate the central and surface angles of the rhombic triacontahedron, let us complete our research into the distances from the centroid to various points of interest on this polyhedron.

We have already calculated the distances to the small and large axis vertices, and to the mid-face. Now we need to find the distance from the centroid to any mid-edge.

Showing the right triangle O'ZA'

We are looking for O'A'. We may write

$\overline{O'A'}^2 = \overline{O'Z}^2 + \overline{ZA'}^2$. We know O'Z, but what is ZA'?

IZ = ½ IJ by construction.

JA' = ½ JV by construction.

The triangles VIJ and ZA'J are congruent by angle-angle-angle. Triangle VIJ is isosceles by construction, therefore triangle ZA'J is isosceles and ZA' = ½rts.

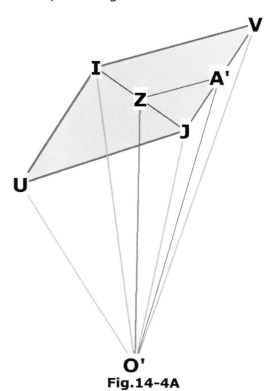

$$\overline{O'A'}^2 = \left(\frac{\Phi^2}{2} * \frac{2}{\sqrt{\Phi^2+1}} \text{rts}\right)^2 + \left(\frac{1}{2}\text{rts}\right)^2 = \left(\frac{\Phi^4}{\Phi^2+1} + \frac{1}{4}\right)\text{rts}^2$$

$$\overline{O'A'} = \sqrt{\frac{\Phi^4}{\Phi^2+1} + \frac{1}{4}}\text{rts} = 1.464386285 \text{ rts.}$$

Fig.14-4A

What are the central angles of the rhombic triacontahedron?

There are 3 central angles of the r.t. that are of interest. Refer to Figure 4.

The first is UO'V, central angle of the long-axis

The second is IO'J, central angle of the short-axis.

The third and primary central angle is IO'V, the central angle of each adjacent side.

O'Z is perpendicular to IJ, and to the plane of the r.t. face, IUJV.

O'Z bisects IJ at Z.

Triangles UO'Z and VO'Z are right.

So we can write $\sin(\angle ZO'V) = ZV / O'V = \dfrac{\left(\dfrac{\Phi}{\sqrt{\Phi^2+1}}\right)}{\Phi}$

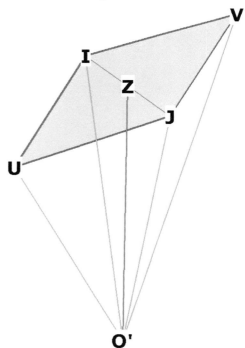

Fig. 14-4, repeated

$$\text{rts} = \frac{1}{\sqrt{\Phi^2 + 1}}$$

$$\angle ZO'V = 31.71747741°$$

We recognize this angle (see Phi Ratio Triangle) as being part of a $1, \Phi, \sqrt{\Phi^2 + 1}$ ratio triangle.

The ratio's of these distances are shown below in Figure 6:

V_____Z_____O'_____V.

      1   Φ $\sqrt{\Phi^2 + 1}$

Figure 6 --- showing the relationship between VZ, O'Z, O'V

Now $\angle UO'V = 2 * \angle ZO'V$, so

$\angle UO'V = 63.4349488°$.

Let's find IO'J, the central angle of the small axis of the r.t. face.

IO'Z and ZO'J are right. So we write

$\tan(\angle IO'Z) = IZ / O'Z.$

We know IZ = one half the side of the dodecahedron, or 1/2 * ds.

From above we know that $O'Z = \dfrac{\Phi^2}{2} ds.$

So $\tan(\angle IO'Z) = \dfrac{\left(\dfrac{1}{2}\right)}{\left(\dfrac{\Phi^2}{2}\right)} = \dfrac{1}{\Phi^2}.$ $\angle IO'Z = 20.90515744°.$

Therefore $\angle IO'J = 2 * \angle IO'Z = 41.81031488°.$

Note that $IZ / O'Z = \dfrac{1}{\Phi^2}$, so that IZ and O'Z have a relationship based on the square of Φ.

|<------------ $\sqrt{\Phi^4 + 1}$ ------------>|

I_____Z_____O'

    1   Φ^2

Figure 7 -- showing the relationship between O'Z, IZ and O'I

This division is in ratio $1, \Phi^2, \sqrt{\Phi^4 + 1}$.

To find IO'V, recognize that the point V lies in a straight line directly above the center of the dodecahedron face. So the angle from IO'V is the same as the angle from I to O' to a point (G) in the middle of the dodecahedron face.

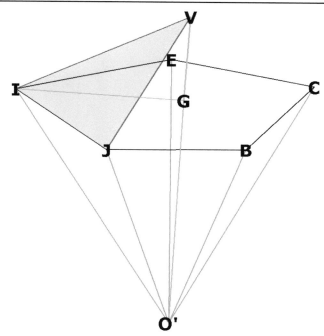

Fig.14-8 shows that the central angle

IO'V = the angle IO'G.

The triangle O'GI is right by construction,

so we need only to know GI and O'G.

We know from *Area of the Pentagon*

that $GI = \dfrac{\Phi}{\sqrt{\Phi^2 + 1}}\, ds$,

and from *Dodecahedron* we know the

distance $O'G = \dfrac{\Phi^3}{2\sqrt{\Phi^2 + 1}}\, ds.$.

Fig. 14-8

So $\tan(\angle IO'G) = GI / O'G = \dfrac{\Phi}{\sqrt{\Phi^2 + 1}} * \dfrac{2\sqrt{\Phi^2 + 1}}{\Phi^3} = \dfrac{2}{\Phi^2}$.

$\angle IO'G = \angle IO'V = 37.37736813°$

What are the surface angles of the rhombic triacontahedron?

Observe from Fig. 14-4 that the angles we are looking for are $\angle IVJ$ and $\angle UIV$.

Z is a bisector of IZ so the triangle IVZ is right.

IV = the side of the r.t. = rts.

$ZV = \dfrac{\Phi}{\sqrt{\Phi^2 + 1}}\, rts$, which we already found above.

So we can write

$\cos(\angle IVZ) = ZV / IV = \dfrac{\Phi}{\sqrt{\Phi^2 + 1}}$.

$\angle IVZ = 31.71747441°$

We recognize immediately, from The Phi Right Triangle , that this angle indicates a right triangle whose long and short sides are divided in Extreme and Mean Ratio (Phi ratio).

Therefore $ZV / IZ = \Phi$.

And so the ratio between the long axis VU and the short axis IJ (see Figure 4) must also be in Mean and Extreme Ratio.

Therefore $VU / IJ = \Phi$.

So the face of the rhombic triacontahedron is a Φ rhombus.

The face angle we want, $\angle IVJ$, is then twice $\angle IVZ$.

$\angle IVJ = 63.43494882° =$ short axis face angle.

One-half of the other face angle, $\angle ZIV$, is just $90 - \angle IVZ = 58.28252558°$.

So $\angle UIV = 116.5650512°$.

What are the lengths of the long axis VU and the short axis IJ of the r.t. face?

$VU = 2 * ZV = \dfrac{2\Phi}{\sqrt{\Phi^2 + 1}} rts = 1.701301617$ rts.

$IJ =$ side dodecahedron $= ds = \dfrac{2}{\sqrt{\Phi^2 + 1}} rts = 1.051462224$ rts.

As stated above,

$$VU / IJ = \dfrac{\dfrac{2\Phi}{\sqrt{\Phi^2 + 1}} rts}{\dfrac{2}{\sqrt{\Phi^2 + 1}} rts} = \dfrac{2\Phi}{2} = \Phi.$$

What is the dihedral angle of the rhombic triacontahedron?

The dihedral angle $= 144°$. For the calculation, see p. 132

What are the relationships between U, Io, Go, O', Ho, and W?

The distance from the centroid to any of the 12 long-axis vertices above the dodecahedron faces $=$

$\Phi \times$ rts.

Distance from centroid to any of the 20 short-axis vertices $= \dfrac{\sqrt{3}\Phi}{\sqrt{\Phi^2 + 1}} rts = 1.473370419$ rts .

We need to go back to Fig. 14-2 and look at the diameter of the sphere which encloses the 12 long-axis vertices of the icosahedron. This line is shown in Fig. 14-2 as UO'W. The diameter passes through the centroid at O' and also through the middle of the 2 large pentagonal planes marked in Fig.14-2.

It also passes through the top and bottom faces of the dodecahedron. These faces are also marked in Figure 2.

There are 4 highlighted pentagonal planes along the diameter UW, as well as the midpoint O'.

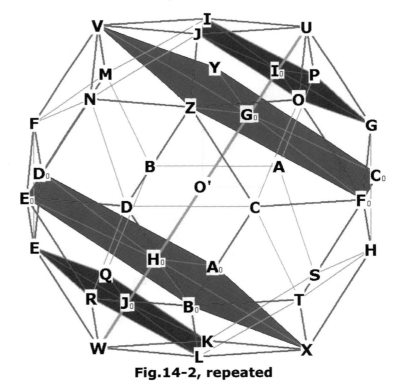

Fig.14-2, repeated

We have the extra distance off the top plane of the dodecahedron to the vertex U, marked as IoU, and the extra distance off the bottom plane of the dodecahedron to the vertex W, marked as JoW.

In Figures 5 and 5A, we found the distance off the plane of the dodecahedron face (IoU , JoW) to be CV.

CV we found to be $\dfrac{\Phi}{\Phi^2 + 1}$ rts.

So IoU = JoW = $\dfrac{\Phi}{\Phi^2 + 1}$ rts.

Refer back to Fig. 14-2.

Let's find the distance UGo, or the distance from U to the first large pentagonal plane VYCoFoZ. This will be easy, because we know that the sides of this pentagonal plane are just the long axes of the r.t. We can form a right triangle from U, to the center Go to any one of the vertices of VYCoFoZ. Let's take the right triangle UGoZ. From Construction of the Pentagon we can get GoZ, it is just the distance from pentagon center to one of the vertices. Inspection of Figure 2 shows that UZ is just the long axis of the r.t. face UJZO.

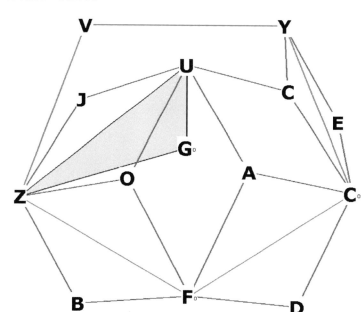

Go is the center of the large internal pentagon. U is directly above Go.
The sides of the large pentagon (in green) are all long axes of the r.t. faces, indicated in red.

Triangle UGoZ is right by construction.

Fig. 14-9 showing the large internal pentagon VYCoFoZ and the right triangle UGoZ.

GoZ $= \dfrac{\Phi}{\sqrt{\Phi^2 + 1}}$ * side of pentagon, or long-axis of r.t. face.

GoZ $= \dfrac{\Phi}{\sqrt{\Phi^2 + 1}} * \dfrac{2\Phi}{\sqrt{\Phi^2 + 1}}$ rts $= \dfrac{2\Phi^2}{\Phi^2 + 1}$ rts.

UZ = side of pentagon, or long-axis of r.t. face $= \dfrac{2\Phi}{\sqrt{\Phi^2 + 1}}$ rts

$\qquad = 1.701301617$ rts.

Now we can find UGo:

$$\overline{UGo}^2 = \overline{UZ}^2 - \overline{GoZ}^2 = \frac{4\Phi^2}{\Phi^2+1} - \frac{4\Phi^4}{(\Phi^2+1)^2}$$

$$= \frac{4\Phi^2(\Phi^2+1) - 4\Phi^4}{(\Phi^2+1)^2}$$

$$= \frac{4\Phi^2}{(\Phi^2+1)^2} = \frac{4\Phi^2}{5\Phi^2}$$

$$= \frac{4}{5}\text{rts.}$$

$$UGo = \frac{2}{\sqrt{5}}\text{rts} = \frac{2\Phi}{\Phi^2+1}\text{rts} = 0.894427191 \text{ rts.}$$

So UGo = 2 * UIo and the plane VYCoFoZ of the dodecahedron is twice the distance from U as is the smaller internal pentagonal plane IPGOJ.

That means UIo = IoGo = 1/2 UGo = $\frac{\Phi}{\Phi^2+1}$rts = 0.447213596 rts.

What is the distance between the centroid O' and the plane IPGOJ, or O'Go?

It is O'U - UGo.

We know O'U, it is the radius of the outer sphere, or $\Phi * \text{rts}$.

Therefore,

$$O'Go = \Phi - \frac{2\Phi}{\Phi^2+1} = \frac{\Phi^3+\Phi-2\Phi}{\Phi^2+1} = \frac{\Phi^3-\Phi}{\Phi^2+1} = \frac{\Phi^2}{\Phi^2+1}\text{rts} = 0.723605798 \text{ rts.}$$

Now we have enough information to make our distance chart of internal planar distances of the rhombic triacontahedron, just as we did for the icosahedron and the dodecahedron.

Distances between internal planes of the rhombic triacontahedron.

Let $\dfrac{\Phi}{\Phi^2+1}=1$.

U_____	_____		_____
1			
Io _____	2	_____	
1			
Go_____	_____		Φ^2+1
		Φ^2	
Φ			
O'_____	_____	_____	_____
Φ			
Ho_____	_____	Φ^2	
1			Φ^2+1
Jo_____	2	_____	
1			
W_____	_____		_____

Table 1.

From the table of relationships we see that:

O'Io is divided in Extreme and Mean Ratio at Go.

O'Jo is divided in Extreme and Mean Ratio at Ho.

UHo is divided in Mean and Extreme Ratio squared at Go.

WGo is divided in Mean and Extreme Ratio squared at Ho.

Conclusions:

The rhombic triacontahedron is a combined icosahedron-dodecahedron dual, so it is not surprising to see so many relationships based on the division in Mean and Extreme Ratio.

The rhombic triacontahedron contains all of the properties of the icosahedron and all of the properties of the dodecahedron and it tells us the proper nesting order of the 5 Platonic Solids.

Dihedral Angle Of The Rhombic Triacontahedron

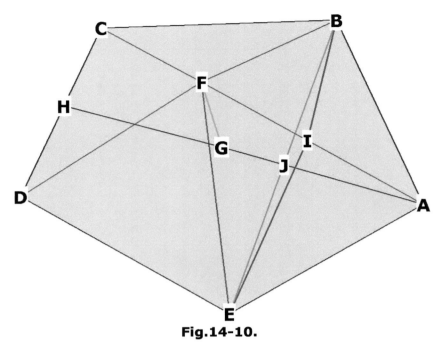

Fig.14-10.

We are looking for the dihedral angle, $\angle$ BIE . Fig. 14-10 shows an r.t. "cap" (in red) over the face of the dodecahedron, and shows that J and G are on AH, the height of the pentagon. F is directly above the mid-face of the pentagon, at G. I is directly above J on BE. $\angle$ BJI is right, allowing us to calculate $\angle$ BIJ or $\angle$ EIJ, and $\angle$ BIE, the dihedral angle, is then just twice that. AH and BE lie on the pentagonal plane ABCDE.

BI is a line along an r.t. face, EI is a line along another of the faces.

Here is our plan of attack:

We know from Rhombic Triacontahedron that $\angle$ BFA is 63.43494882° This can be rewritten

without loss of information as $2 * \sin^{-1}\left(\dfrac{1}{\sqrt{\Phi^2 + 1}}\right)$. Triangle BIF is right by construction. BF is the

side of the r.t., or rts. Here we must use trigonometry to get BI, by taking the sine of $\angle$ BFA so we can find BI. BE is bisected by AH, the angle bisector of $\angle$ BAE and a diagonal of the pentagon. AJ is known from Pentagon Construction. To get $\angle$ BIJ, we can take the sine of $\angle$ BIJ = BJ / BI. The dihedral angle $\angle$ BIE is twice angle BIJ.

First let's find BJ. BJ is one-half the diagonal of the pentagon with sides equal to the side of the dodecahedron. We must change units in terms of the side of the r.t. Remember that

$$ds = \dfrac{2}{\sqrt{\Phi^2 + 1}} \text{rts.}$$

So we write $BJ = \dfrac{1}{2} * \Phi * \dfrac{2}{\sqrt{\Phi^2 + 1}} \text{rts} = \dfrac{\Phi}{\sqrt{\Phi^2 + 1}} \text{rts.}$

$$\sin\left(2 * \sin^{-1}\left(\frac{1}{\sqrt{\Phi^2 + 1}}\right)\right) = BI / BF.$$

$$BI = BF * \sin\left(2 * \sin^{-1}\left(\frac{1}{\sqrt{\Phi^2 + 1}}\right)\right).$$

$$BI = rts * \sin\left(2 * \sin^{-1}\left(\frac{1}{\sqrt{\Phi^2 + 1}}\right)\right) = \frac{2\Phi}{\Phi^2 + 1} rts = 0.894427191 \ rts.$$

Now we can calculate the dihedral angle BIE.

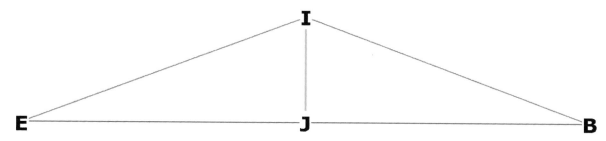

Fig. 14-11 The dihedral angle BIE

We can write

$$\sin(\angle BIJ) = BJ / BI = \frac{\dfrac{\Phi}{\sqrt{\Phi^2 + 1}}}{\dfrac{2\Phi}{\Phi^2 + 1}} = \frac{\sqrt{\Phi^2 + 1}}{2}$$

$\angle BIJ = 72°$.

The dihedral angle BIE is 2 * $\angle BIJ$.

Dihedral Angle = 144°.

Rhombic Triacontahedron Reference Tables

Volume (edge)	Volume in Unit Sphere	Surface Area (edge)	Surface Area in Unit Sphere
12.31073415 s³	2.906170111 r³	26.83281573 s²	10.24922359 r²

Central Angles	Dihedral Angle:	Surface Angle, faces
Long axis: 63.4349488° Short axis: 41.81031488° Adjacent vertex: 37.37736813°	144°	Short axis: 63.43494884° Long axis: 116.5650512

Centroid To: Short Axis Vertex	Centroid To: Mid–Edge	Centroid To: Mid–Face	Side / radius Outer sphere
1.473370419 s	1.464386285 s	1.376381921 s	0.618033989 rts
Long Axis Vertex			Inner sphere (Dodeca-hedron)
1.618033989 s			0.678715947 rts

Chapter 15 – The Nested Platonic Solids

The nested Platonic Solids can be elegantly represented in the Rhombic Triacontahedron, as shown previously.

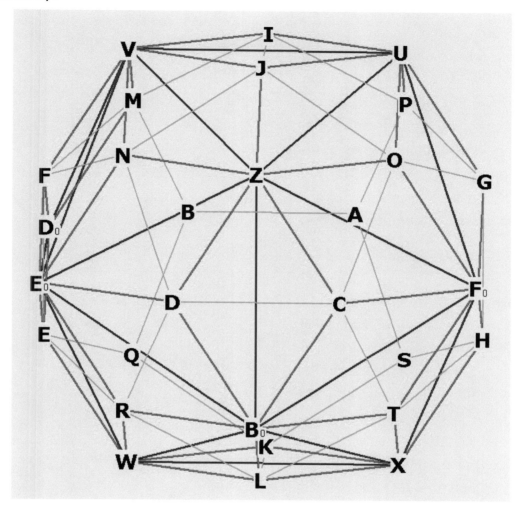

Fig. 15-1: the Rhombic Triacontahedron

Fig. 15-1 shows Phi Ratio rhombi in red, the Icosahedron in blue with its equilateral triangle faces, and the Dodecahedron in black with its pentagonal faces. The Rhombic Triacontahedron is itself a combination of the Icosahedron and the Dodecahedron, and it demonstrates the proper relationship between the 5 nested Platonic Solids. You can see how the edges of the dodecahedron are just the short diagonals of the rhombi of the rhombic triacontahedron, and the edges of the icosahedron are the long diagonals of each of the rhombi of the rhombic triaconta-hedron. Look at ZDB_oC, one of the rhombic faces of the rhombic triacontahedron. Long diagonal ZB is one of the edges of the triangular face ZB_oF_o of the icosahedron, and short diagonal BC is one of the edges of the dodecahedron face JNDCO.

The vertices of the icosahedron are raised off the faces of the dodecahedron, so that the icosa-hedron surrounds the dodecahedron. If you look at point Z, it rises slightly off the center of the dodecahedral face JNDCO.

The cube fits quite nicely within the dodeca-hedron, as shown at right. The cube has 8 vertices and 5 different cubes will fit within the dodecahedron. Each cube has 12 edges, and each edge will be a diagonal of one of the 12 pentagonal faces of the dodecahe-dron. Since there are only 5 diagonals to a pentagon, there can only be 5 different cubes, each of which will be angled 36 de-grees from each other.

(Why is this? Because the diagonals of a pentagon are angled 36 degrees from each other)

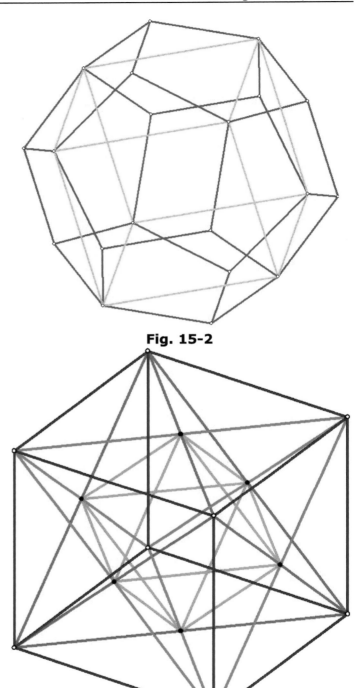

Fig. 15-2

The tetrahedron and the octahedron fit nicely within the cube, as shown above. Fig. 15-3 shows that the octahedron is formed from the intersecting lines of the 2 interlock-ing tetrahedrons. The edges of the tetrahedrons are just the diagonals of the cube faces, and the intersection of the two tetrahedron edges meet precisely at the midpoint of the cube face. If you look at Fig-ure 3 you'll see how the green and purple lines intersect precisely in the middle of the cube face, making an "X." At those points you will see one of the vertices of the octa-hedron.

Fig. 15-3

The nesting order of the 5 Platonic Solids from largest to smallest, as given by the Rhombic Tri-acontahedron, is as follows:

1) Icosahedron
2) Dodecahedron
3) Cube
4) Tetrahedron
5) Octahedron

Here's how the whole thing looks, all enclosed within a sphere:

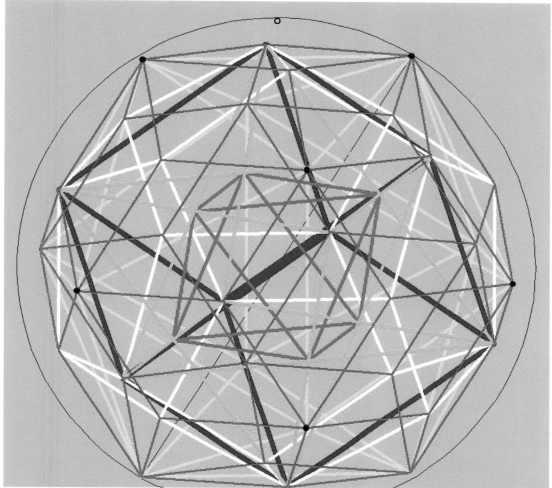

Fig. 15-4: The 5 nested Platonic Solids inside a rhombic triacontahedron, surrounded by a sphere.

The Icosahedron in cream, the rhombic triacontahedron in red, the dodecahedron in white, the cube in blue, 2 interlocking tetrahedra in cyan, and the octahedron in magenta (for those of you viewing in color). Only the 12 vertices of the icosahedron touch the sphere boundary.

Surprisingly, even though there are 5 Platonic solids, there are only 3 different spheres which contain them. That is because the 4 vertices of each tetrahedron are 4 of the 8 cube vertices, and the 8 vertices of the cube are 8 of the 12 vertices of the dodecahedron.

If we let the radius of the sphere that encloses the octahedron = 1, then what is the radius of the other two spheres?

Since the octahedron is formed from the midpoints of all of the cube faces, the sphere which encloses it fits precisely within the cube, like so:

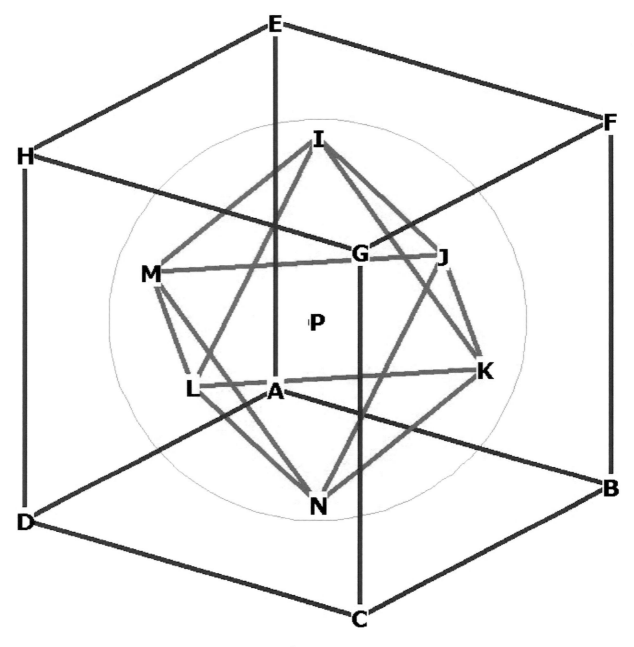

Fig. 15-5

The radius of the circle that encloses the octahedron we will arbitrarily set = 1.

The next largest sphere encloses both the cube and the dodecahedron:

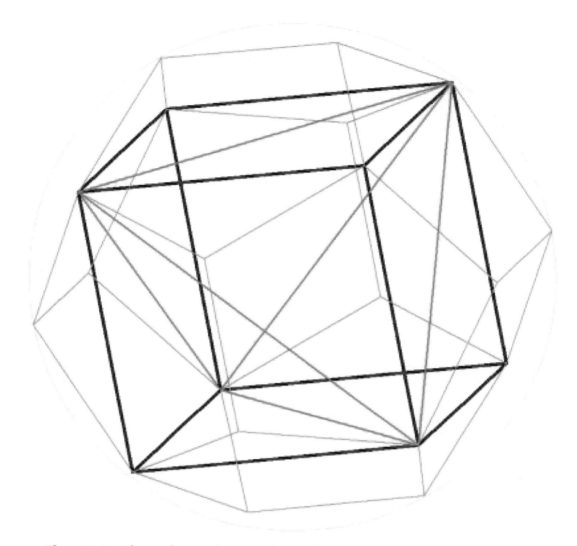

Fig. 15-7: The sphere that encloses both dodecahedron and cube.

The radius of this sphere is $\sqrt{3}$ times the sphere that encloses the octahedron.

The first two spheres are the in-sphere of the cube and the circumsphere of the cube.

The outer sphere that encloses the icosahedron (Figure 4) is slightly larger; $\dfrac{\sqrt{\Phi^2 + 1}}{\sqrt{3}}$ larger, in fact!

So the radii of the three enclosing spheres is: 1, $\sqrt{3}, \sqrt{\Phi^2 + 1}.$

Or, 1, 1.732050808, 1.902113033.

Here is a link to an animated GIF that shows all 5 Platonic Solids and the Rhombic Triacontahedron, rotating inside a sphere: http://www.kjmaclean.com/images/Nested.gif

Appendix A - Combined Reference Charts

Polyhedron	Volume (s³)	Volume (r³)	Surface Area (s²)	Surface Area (r²)
Tetrahedron	0.11785113 s³	0.513200238 r³	1.732050808 s²	4.618802155 r²
Octahedron	0.471404521 s³	1.333333... r³	3.464101615 s²	6.92820323 r²
Cube	1.0 s³	1.539600718 r³	6.0 s²	8.0 r²
Icosahedron	2.181694991s³	2.53615071 r³	8.660254038 s²	9.574541379 r²
Dodecahedron	7.663118963 s³	2.785163863 r³	20.64572881 s²	10.51462224 r²
Cube Octahedron	2.357022604 s³	2.357022604 r³	9.464101615 s²	9.464101615 r²
Rhombic Dodecahedron	8.485281375 r² (distance to 6 octahedral vertices)	3.079201436 s³	2.0 r³ (distance to 6 octahedral vertices)	11.3137085 s²
Icosa Dodecahedron	13.83552595 s³	3.266124627 r³	29.30598285 s²	11.19388937 r²
Rhombic Triacontahedron	12.31073415 s³	2.906170111 r³	26.83281573 s²	10.24922359 r²

Polyhedron	Side / radius
Tetrahedron	0.816496581
Octahedron	1.414213562
Cube	1.154700538
Icosahedron	1.051462224
Dodecahedron	0.713644179
Cube Octahedron	1.0
Rhombic Dodecahedron	To 6 cube vertices: 1.0 To 8 octahedral vertices: 1.154700538
Icosa Dodecahedron	0.618033989
Rhombic Triacontahedron	Outer sphere: 0.618033989 Inner sphere (dodecahedron) 0.678715947

Polyhedron	Central Angle (s)	Dihedral Angle	Surface Angle(s)
Tetrahedron	109.47122064°	70.52877936°	60°
Octahedron	90°	109.4712206°	60°
Cube	70.52877936°	90°	90°
Icosahedron	63.4349488°	138.1896852°	60°
Dodecahedron	41.81031488°	116.5650512°	108°
Cube Octahedron	70.52877936°	125.26438968° (between square and triangular faces)	90°
Rhombic Dodecahedron	Long axis: 90° Short axis: 70.52877936° Adjacent: 54.7356103°	120°	Long axis: 109.4712206° Short axis: 70.52877936°
Icosa Dodecahedron	36°	142.6226318°	60° 108°
Rhombic Triacontahedron	Long axis: 63.4349488° Short axis: 41.81031488° Adjacent vertex: 37.37736813°	144°	Short axis: 63.43494884° Long axis: 116.5650512°

Polyhedron	Centroid to Vertex	Centroid to Mid-Edge	Centroid to Mid-Face	
Tetrahedron	1.0 r	0.577350269 r	0.33333333 r	
	0.612372436 s	0.353553391 s	0.204124145 s	
Octahedron	1.0 r	0.707106781 r	0.577350269 r	
	0.707106781 s	0.5 s	0.408248291 s	
Cube	1.0 r	0.816496581 r	0.577350269 r	
	0.866025404 s	0.707106781 s	0.5 s	
Icosahedron	1.0 r	0.850650809 r	0.794654473 r	
	0.951056517 s	0.809016995 s	0.755761314 s	
Dodecahedron	1.0 r	0.934172359 r	0.794654473 r	
	1.401258539 s	1.309016995 s	1.113516365 s	
Cube Octahedron	1.0 r	0.866025404r	To mid-triangle face: 0.816496581r To mid-square face: 0.707106781r	
	1.0 s	0.866025404s	0.816496581s	0.707106781 s
Rhombic Dodecahedron	1.0 r 1.154700538 s	0.829156198 r	0.707106781 r	
	0.866025404 r 1.0 s	0.957427108 s	0.816496581 s	
Icosa Dodecahedron	1.0 r	0.951056516 r	0.934172359 r	
	1.618033989 s	1.538841769 s	1.511522629 s	
Rhombic Triacontahedron	To short-axis vertex: 1.473370419 s	1.464386285 s	1.376381921 s	
	To long-axis vertex: 1.618033989 s			

Appendix B – The Equilateral Triangle

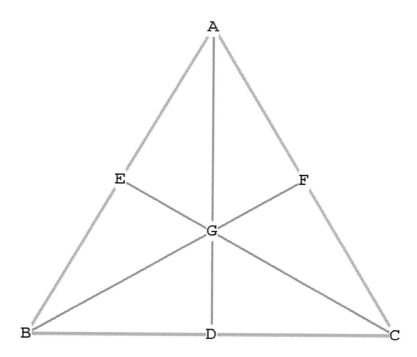

Given an equilateral triangle ABC with side length = s.

Find AG = BG = CG.

Find EG = DG = FG.

Find the height of the triangle (h) AD = BF = EC.

1) AD through G is an angle bisector of <BAC, by construction. AD is a side bisector of BC.

2) CE through G is an angle bisector of <BCA, by construction. CE is a side bisector of AB.

3) BF through G is an angle bisector of <ABC, by construction. BF is a side bisector of AC.

4) triangles AEG, AFG, BEG, CEG, BDG, CDG are all congruent, by side-side side and by construction.

Find h.

By Pythagorean Theorem (PT),

$$h^2 \; = \; \overrightarrow{AC}^2 - \overrightarrow{DC}^2 \; = \; s^2 - \frac{1}{2}s^2 \; = \; s^2 - \frac{1}{4}s^2 \; = \; \frac{3}{4}s^2.$$

$$h \; = \; \frac{\sqrt{3}}{2}s.$$

Find the area of ABC.

The area of any triangle = 1/2 * length of base * height.

The base = CA = s.

$$\text{Area ABC} = \frac{1}{2} * s * \frac{\sqrt{3}}{2}s = \frac{\sqrt{3}}{4}s^2.$$

Show each of the triangles in 4) is similar to triangles ADC and ADB.

5) DC is common to both triangles DCG and DCA..

6) $\angle$ADC and $\angle$CDG are right, and are common to both triangles DCG and DCA.

7) $\angle$DCG = $\angle$DAC because both are bisections of equal angles BCA and BAC.

8) Triangle DCG is similar to triangle DCA by angle-side-angle:

9) Therefore, CG is to CD as AC is to AD, or,

$$CG / CD = AC / AD \quad ---> \quad \frac{CG}{\left(\frac{1}{2}\right)} = \frac{1}{\left(\frac{\sqrt{3}}{2}\right)}$$

$$2 * CG = \frac{2}{\sqrt{3}}.$$

$$CG = AG = BG = \frac{1}{\sqrt{3}} S.$$

$$DG = EG = FG = \frac{\sqrt{3}}{2}S - \frac{1}{\sqrt{3}}S = \frac{3-2}{\sqrt{3}}S = \frac{1}{2\sqrt{3}}S.$$

The longer portion of the diagonal is twice as long as the shorter part:

$$CG / DG = \frac{\left(\frac{1}{\sqrt{3}}\right)}{\left(\frac{1}{2\sqrt{3}}\right)} = 2.$$

Equilateral Triangle Reference Table

Area	Height
0.433012702 s²	0.866025404 s

Center to Vertex	Center to Mid–Edge
0.577350269 s	0.288675135 s

Appendix C – Phi Right Triangles

Right triangles that have sides divided in Mean and Extreme Ratio have the following angles:

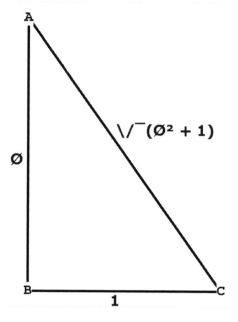

$$\tan(\measuredangle BAC) = \frac{1}{\Phi}$$

$$\sin(\measuredangle BAC) = \frac{1}{\sqrt{\Phi^2 + 1}}$$

$$\cos(\measuredangle BAC) = \frac{\Phi}{\sqrt{\Phi^2 + 1}}$$

$$\measuredangle BAC = 31.71747441°$$

$$\tan(\measuredangle BCA) = \Phi$$

$$\sin(\measuredangle BCA) = \frac{\Phi}{\sqrt{\Phi^2 + 1}}$$

$$\cos(\measuredangle BCA) = \frac{1}{\sqrt{\Phi^2 + 1}}$$

$$\measuredangle BCA = 58.28252559°$$

Appendix D – The Decagon

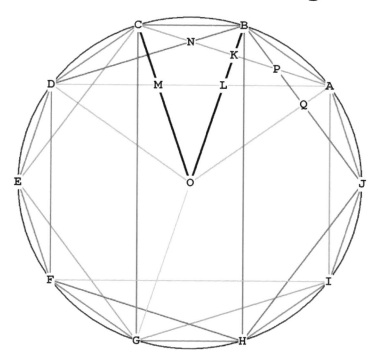

Notice that the decagon is composed of two interlocking pentagons, BDFHJ in purple, and ACEGI, in orange.

We already know the relationship between the radius of the pentagon and the pentagon side, so what is the relationship between the side of the decagon and the radius?

Between the side of the decagon and the side of the pentagon?

We know that OA = OB = OC = OD = radius of circle around decagon/pentagon.

We know, from Construction of the Pentagon Part 2 that OD = OB = radius, is the distance from center to a vertex of the pentagon, and that is $(\Phi/\sqrt{\Phi^2+1}) \times$ side of pentagon.

Inverting this, we know also that side of pentagon = $\dfrac{\sqrt{\Phi^2+1}}{\Phi} *$ radius.

We also know from this paper that OK = $\dfrac{\Phi^2}{2\sqrt{\Phi^2+1}} *$ side of pentagon.

To find AB, the side of the decagon, let's work with triangle BKA.

We know AK to be 1/2 * AC, because OB is a bisector of AC.

$BK = OB - OK = \dfrac{\Phi}{\sqrt{\Phi^2+1}} \times$ pentagon side $- \dfrac{\Phi^2}{2\sqrt{\Phi^2+1}} \times$ pentagon side

$= \dfrac{2\Phi - \Phi^2}{2\sqrt{\Phi^2+1}} = \dfrac{2\Phi - \Phi - 1}{2\sqrt{\Phi^2+1}} =$

$= \dfrac{1}{2\Phi\sqrt{\Phi^2+1}} \times$ side of pentagon.

Then

$$\overline{AB}^2 = \overline{AK}^2 + \overline{BK}^2 = \frac{1}{4} + \frac{1}{4\Phi^2(\Phi^2+1)} = \frac{\Phi^2(\Phi^2+1)+1}{4\Phi^2(\Phi^2+1)}$$

$$= \frac{\Phi^4 + \Phi^2 + 1}{4\Phi^2(\Phi^2+1)} = \frac{4\Phi^2}{4\Phi^2(\Phi^2+1)}$$

$$= \frac{1}{\Phi^2+1}$$

And AB = side of decagon = $\dfrac{1}{\sqrt{\Phi^2+1}}$ * side of pentagon.

Side of pentagon = $(\sqrt{\Phi^2+1})$ × side of decagon.

So side of decagon = $\dfrac{1}{\sqrt{\Phi^2+1}} \times \dfrac{\sqrt{\Phi^2+1}}{\Phi} \times$ radius =

$\dfrac{1}{\Phi} \times$ radius.

$r = \Phi s, \quad s = \dfrac{1}{\Phi} r.$

Therefore the triangle COB, and all of the others like it, are Golden Mean Triangles.

What is DA in terms of the decagon side (ds)? (Pentagon side = ps). See p. 107 and Fig. 13-12.

To get this, we work with triangle DAI.

We know DI, the diameter of the circle surrounding the decagon, or twice the radius.

We also know AI, the side of the pentagon.

So we can write:

$$\overline{DA}^2 = \overline{DI}^2 - \overline{AI}^2 = 4\Phi^2 - (\Phi^2+1) = 3\Phi^2 - 1 = \Phi^4.$$

$\overline{DA} = \Phi^2$ ds.

What is CG = BH? (This is the diagonal of any of the inscribed pentagons).

We will be working with triangle BGH.

We know BG, the diameter or twice the radius, and we know GH = side of decagon.

So $\overline{CG}^2 = \overline{BG}^2 - \overline{GH}^2 = 4\Phi^2 - 1 = \Phi^4 + \Phi^2 = \Phi^2(\Phi^2+1).$

$\overline{CG} = (\Phi\sqrt{\Phi^2+1})$ds.

There are some interesting relationships in the decagon.

Let's start by showing that triangle AOM similar to triangle DOL.

1) $\measuredangle$ DAF = $\measuredangle$ ADI because they subtend equal arcs DF and AI.

2) OD = OA because they are both radii of the circle with center at O.

3) OM = OL because ML is parallel to CB and triangle OML similar to triangle OCB by A-A-A.

Therefore the triangles AOM and DOL are similar by angle-side-side.

Therefore AM = DL.

Since AM = DL and both lines are divided by the equal distance ML, DM = AL.

Now we show that triangles ALO and DMO are congruent.

1) AL = DM

2) DO = AO as they are both radii of the circle centered at O.

3) OM = OL as above.

Therefore the triangles ALO and DMO are congruent by side-side-side.

So AL = OL = DM = OM.

We can write:

LM is to AL as AL is to AM.

Therefore the line AM is divided in Mean and Extreme Ratio at L.

And so the line DL is divided in Mean and Extreme Ratio at M.

ML also divides the line DA in Mean and Extreme Ratio at M and L, so we can also write

AL is to AM as AM is to AD.

These are the same relationships and triangles we saw in the pentagon.

Although we have a lot of information, we still haven't proven the following:

Is AM = MO? We have shown triangles ALO and DMO are congruent, but not isosceles.

This is easily proven, because we know that LM is to MO as CB is to CO, which we have seen, is the division into Mean and Extreme Ratio.

But MO = AL.

Therefore OM is to OC as AL is to AM, and so AM = OC.

So triangle AOM is isosceles.

Since OM is to AM as CB is to OC, then CB = OM.

Therefore triangle AOM congruent to triangle COB.

Also we know that CM = BL = LM.

We know that triangle BAL is a golden mean triangle, since BA = AL, and

BL is to AB as CM is to CB, which is to say, a division in mean and extreme ratio.

∡ CAD = 1/2 that of ∡ BAD, because it subtends an arc exactly 1/2 as long.

Therefore AC is an angle bisector of ∡ BAD, and BK = BL.

****

Let's show all these relationships with brute force calculation!

We will be working first with triangles OKA and AKL.

OB is an angle bisector of <COA, so OB bisects AC at K, and BK = KL.

So OB is perpendicular to AC.

We have previously found BK in relationship to the side of the pentagon. Let's convert this value to the side of the decagon:

$$BK = \frac{1}{2\Phi\sqrt{\Phi^2+1}} * \text{ side of pentagon } =$$

$$= \frac{1}{2\Phi\sqrt{\Phi^2+1}} * \sqrt{\Phi^2+1} * \text{ side of decagon}$$

$$= \frac{1}{2\Phi}\,ds$$

We also found previously that

$$OK = \frac{\Phi^2}{2\sqrt{\Phi^2+1}} * \text{ side of pentagon}$$

$$= \frac{\Phi^2}{2\sqrt{\Phi^2+1}} * \sqrt{\Phi^2+1} * \text{ side of decagon}$$

$$= \frac{\Phi^2}{2}\,ds$$

$AK = $ 1/2 the side of the pentagon, or $\frac{1}{2} \times \sqrt{\Phi^2+1} \times$ side of decagon

$$= \frac{\sqrt{\Phi^2+1}}{2}\,ds.$$

To find AL, write

$$\overline{AL}^2 = \overline{AK}^2 + \overline{KL}^2 = \frac{(\Phi^2+1)}{4} + \frac{1}{4\Phi^2} = \frac{\Phi^2(\Phi^2+1)+1}{4\Phi^2}$$

$$= \frac{\Phi^4+\Phi^2+1}{4\Phi^2}$$

$$= \frac{4\Phi^2}{4\Phi^2}$$

$$= ds.$$

AL = side of the decagon BC, as we stated above.

What is OL? If we are correct above, it should be equal to AL and BC.

OL = OB - BL.

$$BL = BK + KL = 2 * BK = \frac{1}{\Phi}\,ds$$

$$OL = \Phi ds - \frac{1}{\Phi}\,ds = (\Phi - \frac{1}{\Phi})ds = ds.$$

Therefore OL = the side of the decagon = AL = BC.

The relationship of OL to OB is therefore $\dfrac{ds}{\frac{1}{\Phi}\,ds} = \Phi ds.$

We can show DM = AL with the same triangles on the upper left of the decagon.

Since we know DA, we can write:

LM = DA - 2*AL

$$= \Phi^2 ds - 2ds = (\Phi^2 - 2)ds = (\Phi + 1 - 2)ds = (\Phi - 1)ds$$

$$= \frac{1}{\Phi} ds.$$

So by calculation,

$$AM = LM + AL = \frac{1}{\Phi} ds + ds = (1 + \frac{1}{\Phi})ds = \Phi ds,$$

and AM is divided in Mean and Extreme Ratio at L.

What is the relationship between AM and AD. We said it was a Φ relationship. Let's see.

$$AD / AM = \frac{\Phi^2}{\Phi} ds = \Phi.$$

It isn't surprising that there are so many Phi relationships in the decagon, it being composed of 2 interlocking pentagons.

What about AP, BP, KP, and PQ? Are there any Phi relationships there?

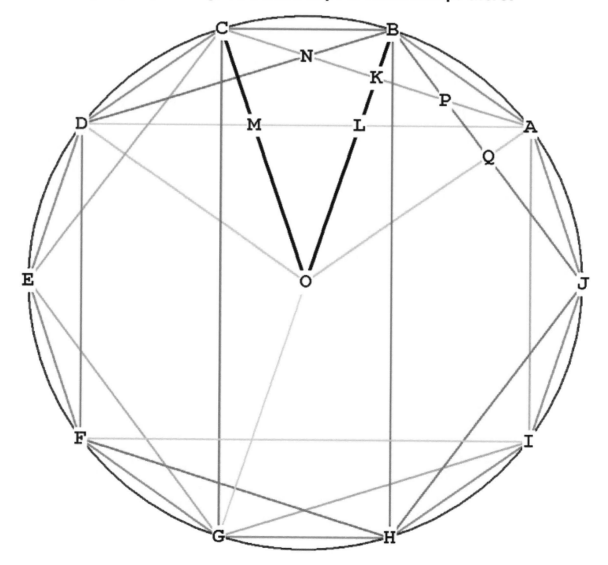

First let's show that triangle BKP is similar to triangle BQO.

1) $\angle$ OBQ is common to both triangles.

2) $\angle$ BKP and $\angle$ BQO are right.

3) Therefore, by the property that all angles of a triangle must add to 180°, $\angle$ BPK = $\angle$ BOA.

So the triangles are similar by angle-angle-angle.

Now we can determine OQ in triangle BQO and from there determine BP and KP.

BQ is 1/2 the side of the pentagon.

$$\overline{OQ}^2 = \overline{OB}^2 - \overline{BQ}^2$$

$$= \Phi^2 - \frac{\Phi^2 + 1}{4}$$

$$= \frac{4\Phi^2 - \Phi^2 - 1}{4}$$

$$= \frac{3\Phi^2 - 1}{4}$$

$$= \frac{\Phi^4}{4}$$

$$\overline{OQ} = \frac{\Phi^2}{2}$$

Now we can write:

BK is to BP as BQ is to OB, or, BK / BP = BQ / OB.

$$BP = OB \times BK / BQ = \frac{\left(\Phi \times \dfrac{1}{2\Phi}\right)}{\left(\dfrac{\sqrt{\Phi^2 + 1}}{2}\right)} = \frac{1}{\sqrt{\Phi^2 + 1}} \, ds.$$

And OQ is to OB as KP is to BP, or, OQ / OB = KP / BP,

$$KP = BP \times OQ / OB = \frac{1}{\sqrt{\Phi^2 + 1}} \times \frac{\left(\dfrac{\Phi^2}{2}\right)}{\Phi} =$$

$$= \frac{\Phi}{2\sqrt{\Phi^2 + 1}} \, ds$$

BN = BP and KN = KP.

The relationship between BP and KP is $\dfrac{2}{\Phi}$.

It looks from the diagram that the sides of the pentagon, BD, BJ, and AC, when they hit each other at N and P, divide the side of the pentagon equally in fourths. But this is not so.

Let's summarize our knowledge of the decagon in the following table:

Table of Data

Identity	Value	Value (decimal representation)
CB = side of decagon	ds	1.0
OB = radius:	Φ ds	1.618033989
AC = side of pentagon:	$\sqrt{\Phi^2 + 1}$ ds	1.902113033
AD = long side of rectangle within decagon:	Φ^2 ds	2.618033989
CG = diagonal of pentagon:	$\Phi\sqrt{\Phi^2 + 1}$ ds	3.077683538
BG = diameter:	2Φ ds	3.236067978
AM	Φ ds	1.618033989
AL = LO = CB	ds	1.0
LM = BL	$\dfrac{1}{\Phi}$ ds	0.618033989
BK	$\dfrac{1}{2\Phi}$ ds	0.309016994
KL	$\dfrac{1}{2\Phi}$ ds	0.309016994
BL	$\dfrac{1}{\Phi}$ ds	0.618033989
BN = BP	$\dfrac{1}{\sqrt{\Phi^2 + 1}}$ ds	0.525731112
KN = KP	$\dfrac{\Phi}{2\sqrt{\Phi^2 + 1}}$ ds	0.425325404

Bibliography

Doczi, G. (1981). *Power of Limits.* Boston: Shambhala.

Densmore, D & Heath, T.L. (2002) Euclid's Elements. Green Lion.

Fuller, R.B. (1975). *Synergetics: Explorations in the Geometry of Thinking* . New York: Scribner

Ghyka, M. (1946) *The Geometry of Art and Life*. New York: Dover Books.

Hart, G.W. *The Encyclopedia of Polyhedra (Virtual Polyhedra)*
> http://www.georgehart.com/virtual-polyhedra/vp.html

Hart, G.W. & Picciotto, H. (2000). *Zome Geometry:* Hands-on Learning with Zome™ Models.
Zome Books.

Herz-Fischler, R. (1998). *A Mathematical History of the Golden Number*. New York: Dover
Books.

Holden, A. (1991). *Shapes, Spaces, and Symmetry*. New York: Dover Books.

Lawlor, R. (1989) *Sacred Geometry: Philosophy and Practice*. London: Thames & Hudson Ltd,

About the Author

Kenneth MacLean is a writer and researcher living in Ann Arbor, Michigan, home of the Michigan Wolverines. He has written six books and dozens of essays.

Contact: kmaclean@ic.net

Website: http://www.kjmaclean.com/